Morten Vollmann

# Combination of Axial Strain Tuning and Gate Bias Tuning for Carbon Nano Tube Resonators

**Morten Vollmann**

# Combination of Axial Strain Tuning and Gate Bias Tuning for Carbon Nano Tube Resonators

Hartung-Gorre Verlag Konstanz

**Reprint of Diss. ETH No. 31252**

**Scientific Reports on Micro and Nanosystems** **Volume 42**

**edited by Prof. Dr. Christofer Hierold**
**ETH Zürich**
**Micro and Nanosystems**

**Cover images:**
**Front:**

- **Left:** Optical image of a wire-bonded MEMS straining platform
- **Middle:** Resonance current increasing with gate bias
- **Right:** Silicon oxide covered MEMS during fabrication

**Back:**

- **Left:** SEM image of the electrodes divided by a 400 nm wide trench
- **Middle:** Gate bias tuning for a suspended carbon nanotube under tension
- **Right:** Visible heat radiation of a thermal actuator

**Bibliographic information published by Die Deutsche Nationalbibliothek**

Die Deutsche Nationalbibliothek lists this publication in the Deutsche Nationalbibliografie; detailed bibliographic data is available in the internet at http://dnb.dnb.de.



**First edition 2025**
**HARTUNG-GORRE VERLAG, KONSTANZ**
**http://www.hartung-gorre.de**

**ISSN 2566-7769**
**ISBN 978-3-86628-855-3**

Diss. ETH No. 31252

# Combination of Axial Strain Tuning and Gate Bias Tuning for Carbon Nano Tube Resonators

A thesis submitted to attain the degree of

DOCTOR OF SCIENCES
(Dr. sc. ETH Zurich)

presented by

Morten Vollmann

born on 16.07.1993

accepted on the recommendation of
Prof. Dr. Christofer Hierold
Prof. Dr. Niels Quack
Prof. Dr. Ashwin Seshia
Prof. Dr. Silvan Schmid

2025

# Abstract

Carbon nanotube (CNT) resonators exhibit exceptional potential for high-frequency and large-range eigenfrequency tuning, yet practical demonstrations have long been limited by the difficulty of precisely applying axial strain. This thesis overcomes that challenge by introducing a custom micro-electro-mechanical system (MEMS) with chevron-style thermal actuators capable of displacing the CNT's clamping ends and generating pure tensile strain levels up to 1.4%. By combining this axial strain with gate-bias-induced strain, frequency tuning of 67 MHz was achieved, equivalent to over 150 % relative tuning, reaching eigenfrequencies over 120 MHz across multiple devices.

To perform such axially straining experiments, new MEMS structures were designed and fabricated according to design restrictions such as the compatibility with dry-CNT transfer methods. Suspended CNT lengths were between 2 and 4.2 µm with gate distances between 490 nm and 2 µm. Different designs of the actuators, thermally heated by electrical current, showed different ability to displace the ends of the electrodes with an efficiency of between 1 and 2.6 $\mathrm{nm\,A^{-2}}$. Including semi-conducting behaviour in FEM simulations explained the displacement response to current not just for low temperatures, but across the full operation range.

Straining experiments on the new MEMS platform included the analysis of the static strain response with gauge factors up to 108 for low gate biases, isolating the axial straining effect from gate bias straining and the combination of both.

A central contribution of this work is the detailed modelling of CNT resonators subject to uniaxial and/or gate bias induced straining. The modelling framework employs solutions to the nonlinear Duffing equation via harmonic balancing, a technique that captures both amplitude and phase behaviour of the resonator's motional current. By including the transfercharacteristic directly into to modelling of the current, even more complex and asymmetric phases and different shapes of resonance peak were explained. This approach proves indispensable for accurately describing the interplay between electrostatic (gate) and mechanical (axial) strain, enabling detailed predictions of device performance. This has been possible only due to the knowledge of applied strain from the characterized actuators. The framework also allowed for the quantification of maximum axial strain of 1.4 % before the CNT slipped from the electrodes.

By validating its design through both experimental measurements and numerical simulations, this thesis demonstrates a viable pathway to high-frequency tunable resonators. The model allows extracting the often unknown pre-strain, which depends on the fabrication circumstances. It also allows investigating subtle effects like mass-loading or effective stiffness. The combination of modelling, engineered device fabrication, and precise straining methods lays the groundwork for future applications in nanoscale sensing or radio-frequency filtering, where ultra-low mass and wide-range tuning capabilities can offer substantial performance gains.

# Zusammenfassung


Resonatoren aus einzelenen Kohlenstoffnanoröhren (CNT-Resonatoren) haben vielversprechende Anwendungsmöglichkeiten, wie etwa im Bereich von Hochfrequenztechnik oder zur Abstimmung von Eigenfrequenzen über ein breites Spektrum. Das Abstimmen der Eigenfrequenzen erfordert ein kontrolliertes axiales Dehnen der CNT, was aufgrund ihrer Abmessungen im Nanobereich bislang nur schwer realisierbar war. Durch die Kombination aus axialer Dehnung und einer transversalen Spannungs-induzierten Dehnung konnten Eigenfrequenzen um bis zu 67 MHz erhöht werden, was einer relativen Erhöhung um über 150 % entspricht. Die gemessenen Eigenfrequenzen lagen bei über 120 MHz auf mehreren Resonatoren.

In dieser Arbeit wird ein mikroelektromechanisches System (MEMS) vorgestellt, um axiale Dehnungsexperimente durchzuführen. Diese MEMS konnten bis zu 1.4 % reine Zugdehnung auf die CNTs wirken. Bei Design und Fertigung der MEMS-Strukuturen wurden Randbedingungen erfolgreich umgesetzt, wie etwa die Kompatibilität mit CNT-Transfermethoden. Die freihängenden CNTs waren zwischen 2 und 4.2 µm lang und hatten Gate-Abstände zwischen 490 nm und 2 µm. Verschiedene Design-Parameter erziehlten unterschiedliche Verschiebungsleistung der Aktuatoren zwischen 1 und 2.6 $\mathrm{nm\,mA^{-2}}$. Das halbleitenden Verhalten vom Silizium wurde in FEM-Simulationen einbezogen und so ließ sich das Verschiebungsverhalten in Abhängigkeit vom angelegten Strom nicht nur bei niedrigen Temperaturen, sondern auch über den gesamten Temperaturbereich hinweg erklären.

Experimente auf der neuen MEMS-Plattform umfassten die Analyse der statischen Reaktion auf Dehnung mit Gauge-Faktoren von bis zu 108 bei niedrigen Gate-Spannungen. Auch Frequenzexperimente konnten nun den Effekt rein axialer Dehnung von gate-bias-induzierter Dehnung trennen, diese Dehnungsarten aber auch kombinieren.

Ein zentraler Beitrag dieser Arbeit liegt in der Modellierung von CNT als hohle Röhren anstatt massiver Zylinder. In der Literatur wurden bisher viele verschiedene Wandstärken zum Modellieren der CNT verwendet.

Durch Experimente und Vergleiche mit molekularen Simulationen zeigt diese Arbeit, dass eine erhöhte Masse der CNTs erforderlich ist, um die beobachteten Eigenschaften von CNT-Resonatoren zu reproduzieren. Die Masse der CNTs lag dabei um Faktoren bis zu siebenfach höher, als von einer makellosen CNT. Diese Ergebnisse konnten mit Messungen über mehrere verschiedene Resonatoren bestätigt werden.

Das Modell basiert auf der nichtlinearen Duffing-Gleichung, welche in dieser Arbeit durch “Harmonic balancing” gelöst wird. Mit diesem Ansatz ist es möglich, sowohl die Amplituden als auch die Phasen der gemessenen Resonanzströme zu erfassen.

Durch die direkte Integration der statischen Übertragungsfunktion in das Strommodell lassen sich zudem komplexere und asymmetrische Phasen sowie unterschiedliche Formen der Resonanz erklären. Dieses Vorgehen is unverzichtbar für eine präzise Beschreibung des Zusammenspiels von elektrostatischer (Gate) und mechanischer (axialer) Dehnung und ermöglicht genauere Vorhersagen zur Geräteperformance. Darüber hinaus konnte mit diesem Modell auch die maximale axiale Zugdehnung von 1.4 % quantifiziert werden, bei welcher die CNT von der Elektrode verrutschte.

Die Aktuatoren wurden anhand von sowohl experimenteller Messungen als auch numerischer Simulationen validiert. So trägt diese Arbeit nicht nur zur Klärung von Materialeigenschaften von CNTs bei, sondern demonstriert auch Strategien zur Realisierung von hochfrequent abstimmbaren Resonatoren.

Die Kombination aus Modellierung, Fertigung und präzisen Dehnungsverfahren bildet die Grundlage für zukünftige Anwendungen in der Nanosensorik oder der Hochfrequenzfilterung, bei denen extrem geringe Massen und ein großer Abstimmbereich erhebliche Leistungsverbesserungen ermöglichen können.

# Acknowledgements

I would like to sincerely thank Prof. Christofer Hierold for everything. His role as my "Doktorvater" has profoundly shaped my development over the past years in many ways. As a mentor and advisor, he intuitively balanced giving me the freedom to explore whatever I was curious about, with stepping in and guiding me more closely whenever I lost direction or got too deep into a rabbit hole.

In scientific discussions, I always felt respected - even when we did not agree and, after months, it turned out he was right. His precision and detail-oriented mindset has fundamentally changed the way I approach problems. His ability to push me beyond what I believed I was capable of is a big part of why I was able to manage this journey.

His calmness and respectful manner, especially during times of deadline pressure or disagreement, will remain a role model for how I aim to handle similar situations in the future. What impressed me most was his deep understanding that every student is different. He adapted his support not only with efficiency, but in a way that was genuinely helpful to each individual.

Secondly, I want to thank Dr. Signe Vehusheia. She has not just been a colleague, team-player, but also partner in both the most darkest and difficult times, when nothing worked in the clean room, and the happiest moments during my PhD. She has been my strongest, most dedicated sparring-partner, my most critical editor and in some ways also coach. I thank her for her support, not just in the work context for reviewing my written texts, but also in my private life. While Prof. Hierold made me a better (and more structured) engineer and scientist, Signe also made me a better human. I could not have completed this journey without you.

The PhD journey is long and often difficult. More often than not, things don't work; the tunnel feels long, dark, and at times, hopeless. At this point, I would like to acknowledge my spiritual guardian. Whenever I felt like giving up, I thought of Thor Heyerdahl. His unwavering dedication and courage to risk his life to prove a scientific hypothesis reminded me that-if he could set sail across the Pacific on a balsa wood raft in the 1940s, from Peru to French Polynesia, and succeed on his first attempt-then I could find the strength to return to the cleanroom yet again, to transfer another carbon nanotube, and then another, even when no "land" was in sight.

I would also like to extend my deepest gratitude to Dr. Cosmin Roman. To this day, it still slightly annoys me that no matter how many weeks I spent researching, solving one problem after another, he would somehow be immediately up to date in our meetings - often knowing the details better than I did. His technical expertise, his intuitive understanding of physics, and his ability to focus energy in the right direction were absolutely essential to the success of this work. He shaped the more technical course of my journey and provided invaluable guidance throughout.

Next, I would like to thank Dr. Miroslav Haluska and Dr. Seoho Jung. Miro, as the CNT expert in the group, consistently provided me with outstanding carbon nanotubes, which were the foundation of much of my work. Seoho, my office partner, welcomed me with open arms and patiently taught me everything I needed to know to get started, both in the lab environment and in working with CNTs. Their support and expertise made it possible for me to begin this journey with confidence.

With regard to the work in the cleanroom, I would especially like to thank Dr. Ute Drechsler, whose support was instrumental in mastering the process flow and overcoming fabrication challenges - even beyond official working hours. She consistently went out of her way to help and took the time to teach with patience and clarity. An hour spent with her often proved more productive than an entire week of working alone. The cleanroom work was also greatly supported by Dr. Antonis Olziersky and Dr. Steffen Reidt, whose expertise and assistance played an important role throughout the fabrication process.

I would also like to extend my sincerest gratitude to the other professors on my examination committee. I first met Prof. Dr. Niels Quack during my master's studies at EPFL, where his "Advanced MEMS" course further increased my enthusiasm for microelectromechanical systems. I am equally grateful to Prof. Dr. Ashwin Seshia, whose thought-provoking discussions at the IEEE MEMS conferences consistently challenged and broadened my perspective. Finally, I owe special thanks to my co-advisor, Prof. Dr. Silvan Schmid, who encouraged me to push the limits of CNT strain by empathy for the struggles during a PhD and provided unwavering support throughout this research.

# List of Symbols

**Table 1:** List of symbols, units and a description.

| Symbol | Unit | Description |
|---|---|---|
| $A$ | $m^2$ | Area of graphene |
| $A_c$ | $m^2$ | Crosssection of the CNT |
| $\alpha_{elec}$ | 1 | Electrode shielding reduction coefficient |
| $\alpha_{ML}$ | 1 | Mass-loading factor |
| $\mathbf{b}$ | m | Initial distance between edge-pairs vector |
| $b_{eff}$ | $kg\,s^{-1}$ | Effective damping ratio |
| $\beta$ | $m^{-1}$ | Wave number |
| $c$ | $m\,s^{-1}$ | speed of light |
| $C_0$ | F | CNT gate capacitance |
| $\bar{C}_g$ | $F\,m^{-1}$ | CNT gate capacitance per length |
| $E$ | Pa | Young's modulus |
| $EA$ | N | Longitudinal rigidity |
| $EI$ | $N\,m^2$ | Bending rigidity |
| $E_{phonon}$ | J | Phonon energy |
| $E_{thermal}$ | J | Thermal energy |
| $\epsilon$ | % | Strain |
| $\epsilon_0$ | % | Pre-strain in CNT |
| $\epsilon_A$ | % | Applied strain |
| $\epsilon_i$ | % | Total strain |
| $F$ | N | Force |
| $f_0$ | Hz | Eigenfrequency |
| $F_{eff}$ | N | Effective excitation force |
| $G$ | $S\,m^{-1}$ | Conductance |
| $G_f$ | 1 | Gauge factor |
| $g_0$ | m | Gate distance |
| $h$ | m | Wall thickness |
| h | $W\,m^{-2}$ | Heat flux |
| $\hbar$ | J s | Reduced Planck's constant |
| $I$ | $m^4$ | Moment of Inertia |
| $i_g$ | A | Conductance modulation current |
| $i_p$ | A | Piezoresistive current |
| $I_{SD}$ | A | Source-drain current |
| $I_{act}$ | A | Actuator current |
| $k_{eff}$ | $N\,m^{-1}$ | Effective stiffness |

| Continuation of Table 1 | | |
|---|---|---|
| $l$ | m | Suspended CNT length |
| $\lambda$ | 1 | Eigenvalue |
| $M$ | N m | Bending moment |
| $m_{eff}$ | kg | Effective mass |
| $\mu$ | $\mathrm{kg\,m^{-1}}$ | Linear mass density |
| $\mu_e(\mathrm{T})$ | $\mathrm{cm^2\,V^{-1}\,s^{-1}}$ | Electron mobility |
| $\mu_h(\mathrm{T})$ | $\mathrm{cm^2\,V^{-1}\,s^{-1}}$ | Hole mobility |
| $n_i$ | $\mathrm{cm^{-3}}$ | Intrinsic carrier concentration |
| $n_{phonon}$ | 1 | Phonon number |
| $N_h$ | 1 | Truncation order |
| $\phi_n(x)$ | 1 | Mode shape for eigenmode $n$ |
| $Q$ | 1 | Quality factor |
| Q | N | Shear force |
| $q_c$ | C | Charge |
| $r$ | m | CNT radius |
| $R$ | Ω | Resistance |
| $R_0$ | Ω | Initial resistance |
| $R_{act}$ | Ω | Resistance of a thermal actuator |
| $\rho$ | $\mathrm{kg\,m^{-2}}$ | Areal mass density |
| $\sigma_r$ | $\mathrm{m^2}$ | Radial displacement variance |
| $t$ | s | Time |
| T | K | Temperature |
| $T$ | N | Tension |
| $T_{ax}$ | N | Axially applied tension |
| $T_g$ | N | Gate bias induced tension |
| $T_0$ | N | Pre-tension |
| $V_{GD}$ | V | Gate voltage |
| $V_{SD}$ | V | Source voltage |
| $z_s$ | m | Static deflection |
| $z_{dn}$ | m | Dynamic amplitude of $n^{th}$ order |

# List of Abbreviations

**Table 1:** List of abbreviations and a description.

| Abbreviation | Description |
|---|---|
| AC | Alternating current |
| ALD | Atomic layer deposition |
| AlN | Aluminium Nitride |
| BAW | Bulk acoustic wave |
| BHF | Buffered hydrofluoric acid |
| CMR | Contour mode resonators |
| CNT | Carbon nanotube |
| Cr | Chromium |
| CVD | Chemical vapour deposition |
| DC | Direct current |
| DMSO | Dimethyl Sulfoxide |
| DRIE | Deep-reactive ion etching |
| FEM | Finite element modelling |
| FET | Field effect transistor |
| GF | Gauge factors |
| GPIB | General purpose interface bus |
| GaN | Gallium Nitride |
| HB | Harmonic balancing |
| h-BN | hexagonal Boron Nitride |
| IPA | Isopropanol |
| LoG | Laplacian of Gaussian |
| LPCVD | Low-pressure chemical vapour deposition |
| MEMS | Micro-electro-mechanical system |
| MMA | Methyl Methacrylate |
| MIBK | Methyl Isobutyl Ketone |
| $MoS_2$ | Molybdenum Disulfide |
| MWCNT | Multi walled carbon nanotube |
| $NO_2$ | Nitrogen Oxide |
| PCB | Printed circuit board |
| Pd | Palladium |
| PMMA | Polymethyl Methacrylate |
| PZT | Lead Zirconate Titanate |
| RBM | Radial breathing mode |
| RIE | Reactive ion etching |

| Continuation of Table 1 | |
|---|---|
| SAW | Surface acoustic wave resonator |
| SEM | Scanning electron microscope |
| SGS | Small band gap semiconducting |
| SiN or $Si_xN_y$ | Silicon Nitride |
| $SiO_2$ | Silicon Oxide |
| SMR | Solidly mounted resonators |
| TIA | Trans impedance amplifier |
| UV | Ultraviolet |
| WLI | White light interferometer |
| ZnO | Zinc Oxide |

# Contents

# List of Figures

# List of Tables

# 1 Introduction

The first tunable resonator might have been a string instrument like the lyres or harps in Mesopotamia already 2600 BCE [1] (Fig. 1.1a). The stringed instrument required a mechanism to tighten or loosen the tension of the string in order to tune the tone (or eigenfrequency) of each string in order to achieve harmonic sounds. String instruments were developed all over the world and the tuning mechanism of axial strain applied through the tuning pegs remained simple and precise. Even today they are still used in violins, pianos and guitars. The importance of eigenfrequency tuning becomes immediately evident once one string is out of tune.

It took millennia before a second notable application of tunable resonators was discovered by the florentine physicist and astronomer Galileo Galilei in 1582 [2]. He realized that the swinging frequency of a pendulum depends only of the square root of its length. It took until 1656 until the dutch scientist Christiaan Huygens used this tunable resonator in an application [3]. The pendulum clock increased the precision of time keeping from several minutes a day down to just a few seconds and remained the standard until the 1930's.

A third example of a tuned resonator is found in civil engineering. The resonance catastrophe of the Tacoma Narrows Bridge (1940) is often used in mechanics lectures all over the globe. The bridge collapsed because strong winds hit it transversally and triggered large vibrational amplitudes in its eigenfrequency until it fell apart. A more unpredictable excitation of oscillation in civil engineering are seismic waves in the form of earthquakes. To prevent structural failure of the Taipei 101 in Taiwan, a tunable resonator is installed on the top floors of the building (Fig. 1.1b). The eigenfrequency is tuned to match the eigenfrequency of the building, then in cases of strong winds or earthquakes, the oscillation of the coupled resonator is damped and thereby the overall vibration is attenuated.

Now that the relevance and diversity of tunable resonators in the history of the macroscopic world is covered, lets take a look into the development of resonators on the micro scale. In the 1950s the first quartz crystal resonators were used in electronics for frequency control in radios or early computing systems. These resonators already offered high precision. In 1965, with the establishment of Moore's Law as a roadmap for the semiconducting industry, the doubling of transistors on a chip every 2 years

**(a)** Reconstructed “Queens lyre” from 2600 BC with tuning pegs [1]. ©The Trustees of the British Museum. Shared under a Creative Commons Attribution - NonCommercial - ShareAlike 4.0 International (CC BY-NC-SA 4.0) licence.

**(b)** Spring mass damping system spanning over the top floor of Taipeh 101, Taiwan's highest building (508 m). Picture taken during trip for IEEE MEMS 2025 in Taiwan.

**Figure 1.1:** Real world examples of tunable resonators.

was predicted [4], resonators needed to keep up with further miniaturization to replace bulkier components. Microtechnology boomed, and in 1979 the first resonators were used in RF communication as filters and only a year later the first microelectromechanical system (MEMS) resonators were developed and integrated into electronic circuits. MEMS technology evolved further with advances in materials such as piezoelectric MEMS with aluminum nitride or zinc oxide. Resonating MEMS began to be used commercially in acceleration sensors, pressure sensors or microphones. [5–7]

With the first report of carbon nanotubes (CNTs) by Ijima in 1991 [8] and the discovery of graphene in 2004 [9] (2010 Nobel prize in Physics), two materials were presented, which would drag the focus of resonator research onto them. Both materials have been object of relentless research interest in tunable resonators at the micro and nanoscale. Various architectures for both one- and two-dimensional resonators were explored in order to tune their properties such as eigenfrequency.

In this thesis, tunable CNT resonators are studied. CNTs show exceptionally high yield strain, while combining high stiffness and low mass, making them a promising candidate for tunable resonators.

# 2 State of the Art

In this chapter, an overview over different resonator types is given and their potential for frequency tuning is explained. A state of the art of the achieved tuning ranges and methods is given. At the end of this chapter, the structure of the thesis is outlined.

## 2.1 The Family of Micro and Nano Resonators

MEMS resonators can be separated into three categories based on their main functional dimensions: 3D (bulk), 2D (membrane), and 1D (beam). In this section, an overview of the different architecture types is given and their applicability for eigenfrequency tuning is discussed in relation to current literature and research.

Resonators are characterized by their eigenfrequency, often denoted as $f_0$, describing the frequency of largest amplitude response and quality factor $Q$ which describes the width of that response. In this section, the *tunability* or *tuning range* of resonators is compared, which is a measure of how much the eigenfrequency can be shifted.

### 2.1.1 3D Bulk Resonators

3D bulk resonators usually work as continuum resonators and consist of standing elastic waves in their oscillating body. They are usually made of piezoelectric materials. Three types of 3D bulk resonators are discussed here: bulk acoustic wave (BAW) resonators, surface acoustic wave resonators (SAW), and contour mode resonators (CMR). [10, 11]

BAW resonators can handle high power, achieve large eigenfrequencies and high quality factors. They are often used as filters and can be further divided into film bulk acoustic resonators (FBARs) and solidly mounted resonators (SMR). In both designs, the resonator consists of a piezoelectric material such as aluminum nitride (AlN) [12], scandium doped aluminum nitride (AlScN) [13], lithium niobate (LN) [14, 15], gallium nitride (GaN), zinc oxide (ZnO) [16], and lead zirconate titanate (PZT) [17] with a top and bottom electrode. A schematic is shown in Figure 2.1a. Alternating voltages are applied onto these electrodes which create an acoustic wave. This causes a resonance in the structure. Electrode spacing into multiples/fractions of the elastic wavelength is crucial for actuation efficiency. The eigenfrequencies of such devices ranges from MHz [18] up to several tens of GHz [17] and is defined by the material parameters such as stiffness, density, poissons ratio, acoustic velocity, acoustic impedance, dielectric constant, and the thickness of the bulk. While BAW resonators have high $Q$ and high frequencies, it is difficult to fabricate them with larger bandwidth and they do not exhibit promising angles for frequency tuning [19]. BAW resonators have been tuned before e.g. from 5.45 GHz down to 5.3 GHz by applying 50 V which triggered electrostatic spring softening with quality factors around 250 [20]. The equations for electrostatic spring softening are derived in Section 3.3.2.

A similar working principle to the BAW is the SAW, however the travelling direction of the acoustic/elastic wave is not into the bulk, but along its surface (Fig. 2.1b). They have been used as filters in the first three generations of mobile (cellular) communication networks with frequencies between 2 GHz and 850 MHz [21]. These filters play a crucial role in signal processing, enabling efficient frequency selection and reducing interference in wireless communication systems. The quality factors reach around 1000 [22, 23] at ambient pressure and can reach over half a million at low pressures and low temperatures [24].

CMRs build the next member of the bulk resonator family. Here, the eigenfrequency is again determined by the geometry of the resonator, however not by the thickness of the resonator, but by the in-plane lateral geometry. This allows for the design of multiple resonators on the same chip with different eigenfrequencies. [25–29]. CMRs can be used in multi-band RF filter applications.

The 3D class of bulk resonators does not exhibit large potential for frequency tuning, because the eigenfrequency is mostly determined by resonator geometry or material parameters.

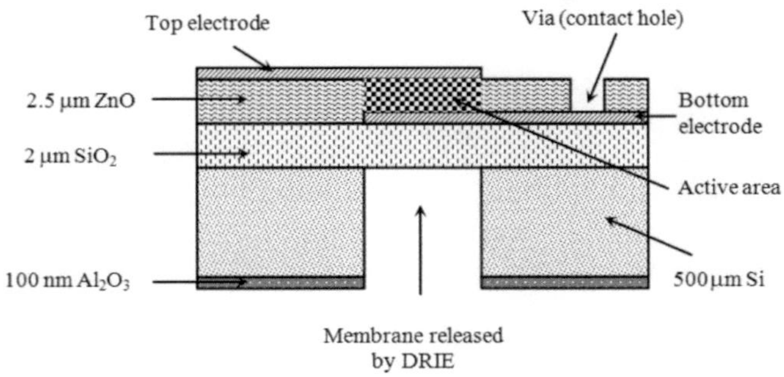


**(a)** Schematic cross-section of a Film Bulk Acoustic Resonator (FBAR) [16]. The device consists of a thin piezoelectric ZnO film sandwiched between two electrodes, forming the resonant structure. The active area is released from the underlying silicon using Deep Reactive Ion Etching (DRIE) to form a suspended membrane, allowing acoustic waves to resonate in the vertical direction. The multilayer stack includes silicon oxide and alumina layers for insulation and mechanical support. ©2011 IEEE.

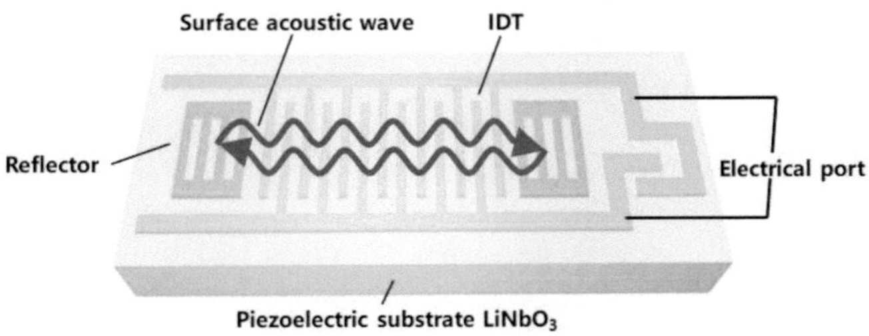


**(b)** Working principle of a Surface Acoustic Wave (SAW) resonator (adapted from [22]). Surface acoustic waves are generated and detected using interdigitated transducers (IDTs) patterned on a piezoelectric substrate. Reflectors on either side of the IDT help confine the wave within the active region. Electrical signals applied to the IDTs are converted to mechanical waves via the piezoelectric effect. Open Access figure from [22] published under open access under CC BY licence (Creative Commons Attribution 4.0 International license).

**Figure 2.1:** Examples of bulk acoustic wave (BAW) resonators. (a) illustrates a vertical FBAR resonator based on membrane release technology for high-frequency operation. (b) shows a SAW resonator utilizing surface-propagating waves confined by reflectors on a piezoelectric substrate. Both are widely used in RF filters and sensing applications due to their compact size and high $Q$-factors.

### 2.1.2 2D Resonators

Unlike 3D resonators, which have their resonance modes in the bulk material, 2D resonators exhibit flexural modes. These structures are thin in one dimension, resembling a sheet. The category of 2D resonators includes plate, membrane, and trampoline resonators. While plate and membrane resonators share similar architectures, their mechanical behaviour is influenced not only by their geometry and material properties but also by tensile stress. When the tensile stress is low and bending stiffness is dominating, we refer to them as plate resonators, and when the tensile stress is high, they are called membrane resonators.

The membrane resonator consists of a thin suspended film under tension. The tension dominated dynamics make them highly sensitive to external pressure changes. Since the eigenfrequency is tension dependent, these resonators can be frequency tuned. Thin membranes consist of materials such as silicon nitride, achieving frequencies of up to several MHz [30]. With advances in novel materials, these membranes can now consist of monolayer structures of hexagonal boron nitride (h-BN) [31], black phosphorous [32], molybdenum disulfide ($MoS_2$, see Fig. 2.2a) [33, 34] or even graphene [35].

Trampoline resonators combine elements of both plate and membrane resonators. They consist of a 2D central oscillating region that is suspended by narrow and compliant tethers (Fig. 2.2b). These structures reduce anchor loss effects and therefore usually exhibit higher $Q$ factors of hundreds [36, 37] up to multiple million [38, 39] in comparison to other 2D monolayer resonators such as graphene or $MoS_2$ which usually remain far below 1000 in vacuum and ambient pressure. Applications for trampoline resonators are currently in the field of optomechanics [37, 38].

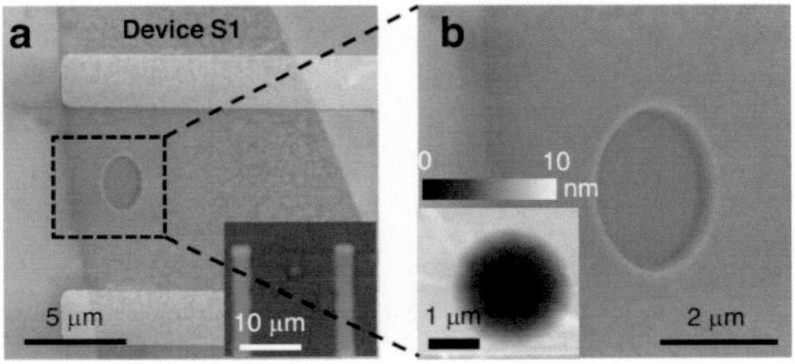


**(a)** Suspended $MoS_2$ membrane resonator. The monolayer $MoS_2$ forms a circular suspended region of 2 µm diameter. The light gray areas indicate source and drain electrodes. Inset in (b) shows an AFM image of the suspended membrane with measured deflection. Open Access figure from [40] published under open access under CC BY licence (Creative Commons Attribution 4.0 International license).

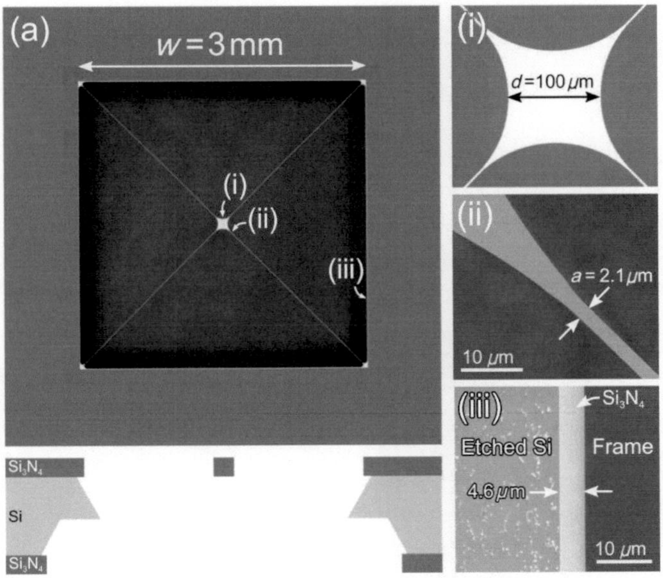


**(b)** Example of a 2D trampoline resonator structure from [38]. The silicon nitride trampoline area is visible in (i), the 4 tethers are closed up in (ii). The schematic cross section is visible in (a). ©2016 American Physical Society.

**Figure 2.2:** Examples of 2D resonators. (a) illustrates a $MoS_2$-based membrane resonator, and (b) shows a large-scale silicon nitride trampoline resonator with tethers.

### 2.1.3 1D Beam Resonators

1D beam resonators are suspended structures which have two dimension that are much shorter than its last. These 1D nanostructures exhibit different mechanical dynamics depending on its tensile stress state. At low stresses the bending stiffness is dominating and the structure behaves like a beam. At higher tension the structures behave more like a string (similar to plates vs. membranes). The resonance behaviour of these structures further depends on the boundary conditions of the system. Beam usually refers to "clamped"-"clamped" and cantilever refers to "clamped"-"free".

Examples of 1D resonators include nanowires (NW) which are solid structures consisting of silicon [41–43] (Fig. 2.3a), silicon nitride [44], silicon carbide [45], gallium nitride [46] (Fig. 2.3b) or platinum [47]. NW have a low mass because of their small dimensions, they exhibit high stiffness and therefore have high eigenfrequencies of up 200 MHz in their fundamental mode. Since these structures are brittle and have a low yield strain, they are not suitable for achieving large tuning ranges as they would break.

Further, CNTs can also be used as material for 1D resonators. They exhibit high stiffness ($E = 1\,\mathrm{TPa}$, [48]), low mass ($\mu < 5 \times 10^{-15}\,\mathrm{kg\,m^{-1}}$, [49]) and high elasticity (strain limit $\epsilon_{max} \approx 6\,\%$, [50]). On top of these outstanding mechanical properties, CNTs have remarkable conductive properties, such as electrical conductance up to $10^{7}\,\mathrm{S\,m^{-1}}$ [51] and thermal conductivity of over $2000\,\mathrm{W\,m^{-1}\,K^{-1}}$ [52]. Specifically the high yield strain of CNTs make them the perfect material for large tuning ranges. Tuning approaches to exploit the unique combination of parameters are discussed in the next section.

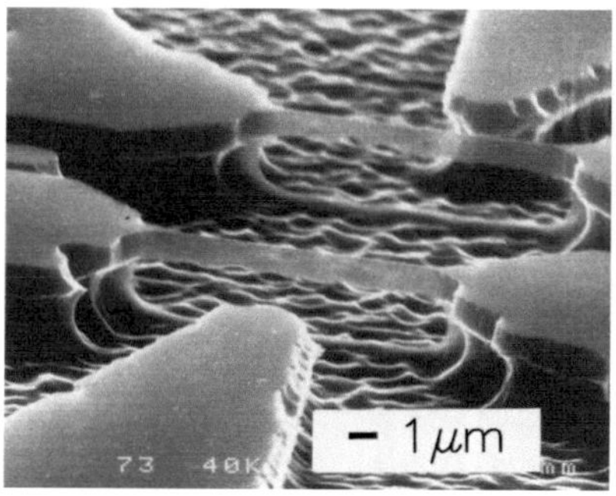


**(a)** Two 0.8 µm high silicon nanowires from bulk silicon forming two doubly clamped beam resonators [42]. ©1996 AIP Publishing.

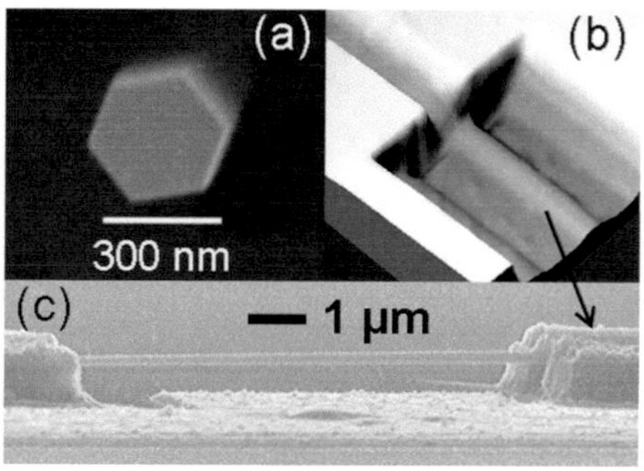


**(b)** GaN nanowire resonator with hexagonal cross-section (a), an AFM image of the suspended wire (b), and an SEM image showing the suspension (c). Adapted from [46]. ©2011 AIP Publishing.

**Figure 2.3:** Examples of 1D resonators.

## 2.2 Tuning Mechanisms of Doubly Clamped 1D Resonators

The most promising design for tunable 1D beam resonators consists of a doubly clamped suspended CNT in a field effect transistor (FET) configuration. Its frequency is determined by the combined continuum dynamics of a beam-string, whereas the beam behaviour is dominant for low stress [53]. The eigenfrequency $f_0$ of 1D doubly clamped resonators is as follows [54]:

$$f_{0,n} = \frac{1}{2\pi}\sqrt{\left(\frac{\lambda_n^2}{l^2}\cdot\sqrt{\frac{EI_y}{\mu}}\right)^2 + \left(\frac{n\cdot\pi}{l}\cdot\sqrt{\frac{\epsilon\cdot EA_c}{\mu}}\right)^2} \tag{2.1}$$

where $\lambda_n$ is the eigenvalue of the $n^{th}$ mode, $l$ is the suspended CNT length, $E$ is the Young's modulus, $I_y$ the moment of inertia, $\mu$ the linear density, $A_c$ is the cross-sectional area and $\epsilon$ is the strain. A more detailed derivation and discussion about the mechanics is given in Chapter 3 (compare Eqn. 3.15), but here this equation is highlighted to understand the important influence of strain $\epsilon$ onto the eigenfrequency of the resonator. To influence or *tune* the eigenfrequency, strain needs to be applied.

### 2.2.1 Gate Bias Tuning

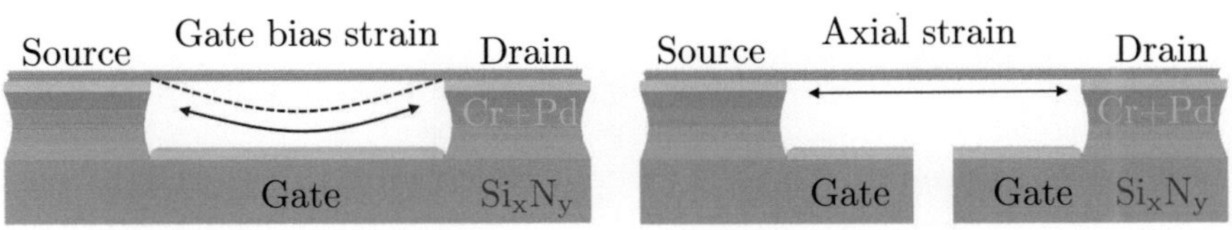


**(a)** Gate bias tuning mechanism. **(b)** Axial strain tuning mechanism.

**Figure 2.4:** Schematic of the two strain tuning mechanisms for CNT resonators.

The most common tuning mechanism is *gate bias tuning* (see Fig. 2.4a), where an electrostatic force of the gate electrode pulls the CNT closer and thereby induces strain. This method is easily implemented, as it requires no further design features since the gate electrode is already part of the design. Gate bias tuning was the first demonstrated CNT tuning mechanism with tuning ranges already up to 330 % [55]. Since then, gate bias tuning has been widely studied for mass responsivity down to 11 Hz yg$^{-1}$ to evaporated chromium atoms [56], for $Q$-factor improvement at low temperatures [57], improved modelling [58, 59], and nonlinear spring hardening and softening [57, 60].

As seen in Equation 2.1, the eigenfrequency $f_0$ is not just dependent on the CNT strain, but also on its suspended length $l$. Many resonators have been demonstrated with suspended length $l$ of a few microns achieving eigenfrequencies of several tens of MHz [54, 55, 61, 62]. Shorter CNT resonators with suspended lengths in the range of hundreds of nanometers have been explored, for example 150 nm ($f_0$ = 1.84 GHz, [63]), 161 nm ($f_0$ = 2.08 GHz, [64]), or 300 nm ($f_0$ = 280 MHz, [65]). The eigenfrequencies here do not purely correlate directly with Equation 2.1, because in contrast to the well defined length, the initial pre-strain is not known. The pre-strain is difficult to be controlled during the fabrication process or by the device architecture. Some CNTs even exhibit slack upon fabrication (like a hanging chain).

Slack modes in CNT resonators have been investigated beyond the standard electrode layout. More complicated back and side gates have been introduced. These were able to not just trigger hardening effects (eigenfrequency shifts up), but also a reduction of eigenfrequency (softening) [60].

*Gate bias tuning* is inherently limited by the applicable voltage onto the gate electrode until excessive gate bias leads to electrical break-down or triggers the pull-in instability. This limit poses a problem of this method for potential device applications, because the voltage necessary for desired tuning ranges might not be available and the gate bias should remain reserved for controlling the transistor response. The gate bias tuning method has virtually no power consumption, because no current is flowing. A few realized designs are presented in Figure 2.5.

Gate-tunable CNT resonators can be fabricated using two distinct methodologies: the "bottom-up" and "top-down" approaches. In the "bottom-up" approach, the CNT resonators were fabricated by locating a CNT grown by chemical vapor deposition (CVD) on a substrate. Then the ends of the CNT are contacted with metal such as gold in an e-beam deposition process. The CNT is then underetched to define the suspended length [55, 58]. Two examples can be seen in Figure 2.5a and 2.5b. In Figure 2.5a top, slack can be seen and is indicated in the sketch below. Both approaches end up to have a "top-clamping" configuration.

The second approach fabricates the device substrate and the CNT separately. The MEMS structure on which the CNT is placed (also called the receiving MEMS substrate) is fabricated using silicon micromachining, while the CNT is grown with CVD on a separate growth substrate. The CNT is then transferred using a dry transfer process onto the receiving MEMS substrate, resulting in an adhesively bonded clamping

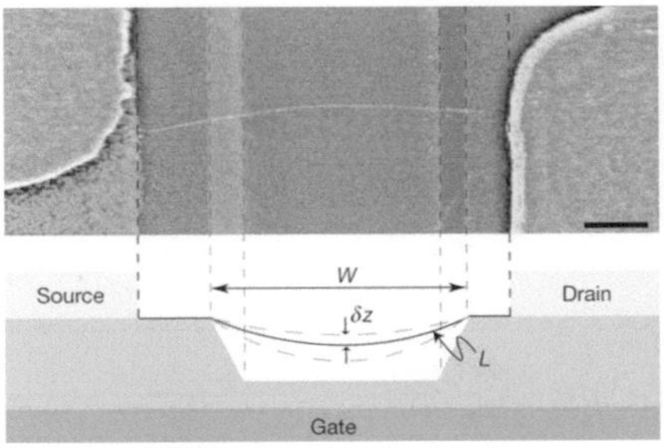


**(a)** First presented CNT resonator with visible slack. Schematic overlay illustrates the suspended CNT geometry and shows the top-clamping configuration [55]. ©2004 Springer Nature.

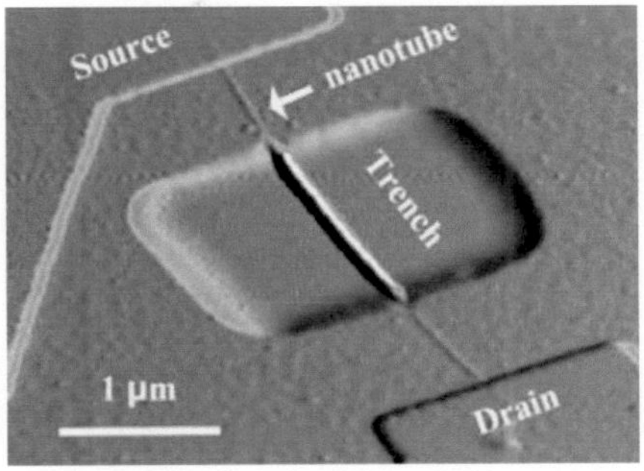


**(b)** AFM image of a similar CNT resonator design showing the suspended nanotube across a trench [59]. ©2007 WILEY-VCH.

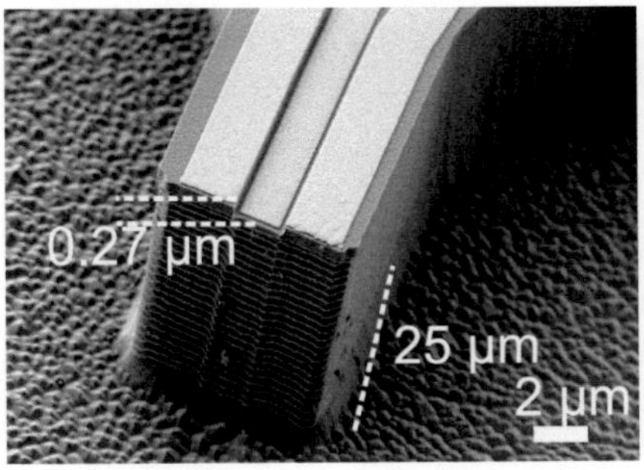


**(c)** SEM image from previous work in the Micro- and Nanosystems group, showing a mesa-etched platform used for dry transfer of nanotubes. Open Access figure from [66] published under open access under CC BY licence (Creative Commons Attribution 4.0 International license).

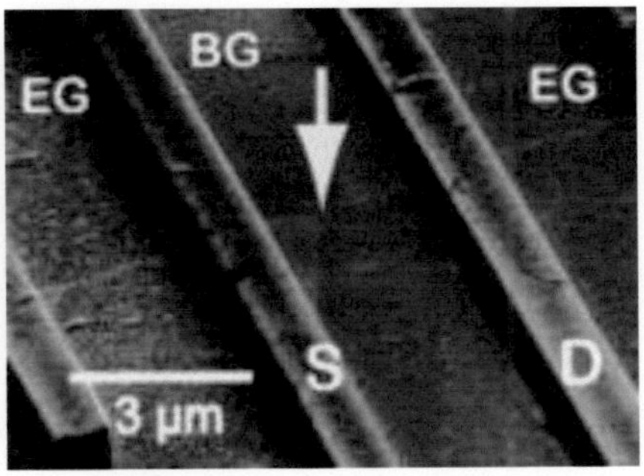


**(d)** Long suspended CNT with a complex multi-electrode structure. EG = edge gate, BG = back gate, S = source, D = drain. Reprinted with permission from [60]. ©2011 American Chemical Society.

**Figure 2.5:** Overview of different gate tuning mechanisms for CNT resonators. The images illustrate a progression of design concepts. Early devices with top clamping are shown in (a) and (b), with visible slack in (a). A second fabrication approach based on dry transfer using mesa structures is shown in (c), while (d) demonstrates complex electrode geometries that enable more sophisticated tunable electrostatic control.

configuration. There are multiple different approaches for the dry transfer [60, 67–69]. Figure 2.5c shows a receiving substrate with a 25 µm deep mesa etch for CNT resonators or gas sensors from previous work in the Micro- and Nanosystems group [66]. Figure 2.5d shows a different substrate with longer CNTs and a more complex electrode structure. A closer discussion on the fabrication of the CNT and MEMS used in this thesis are presented in chapter 4.

### 2.2.2 Axial Strain Tuning

More effective than the gate bias tuning is the *axial strain tuning* (see Fig. 2.4b), where the straining force is not applied orthogonally to the direction of the strain, but directly in the strain direction. MEMS straining platforms capable of applying this tuning method are also able to remove initial slack, a fundamental advantage. Slack modes are less stable and less predictable and therefore undesired.

In contrast to gate bias tuning, axial strain tuning has not been studied as extensively for CNTs, because the required setups are much more sophisticated in terms of design and fabrication. Therefore, the examples for axial strain tuning are scarce.

The most successful system in terms of achieved eigenfrequencies and tuning range is a setup presented by Ning et al. in 2014 [70]. They placed a CNT with nanomanipulators over a slit in the substrate (Fig. 2.6a). The source electrode was placed on a piezoelectric stage, allowing it to move precisely (0.016 nm), while the whole setup was placed in an SEM chamber, allowing for direct measurement of displacement and therefore applied strain. They were able to tune CNT eigenfrequency from 21.04 MHz up to 430.62 MHz, achieving strain levels of up to 2 %. This represents the highest achieved tuning range for resonators to date. However, while the piezoceramic stage presents an excellent platform for CNT resonator research, it is bulky and not suitable for potential future applications. Therefore, other straining platforms have been investigated using MEMS technology to reduce the form-factor, facilitate easier readout and actuation.

One notable approach for a straining platform was presented by Goldsche et al. in 2018 [71], however, it was not used for mechanical resonance, but only strain experiments. They used an electrostatically actuated comb structure to strain graphene (Fig. 2.6b). A static applied voltage between the fingers of the combs creates an attractive force and thereby moves the structure against the restoring spring forces. They achieved controlled applied displacement of up to 60 nm with forces up to 200 nN at 3 V applied over their capacitive actuators. This MEMS approach offers

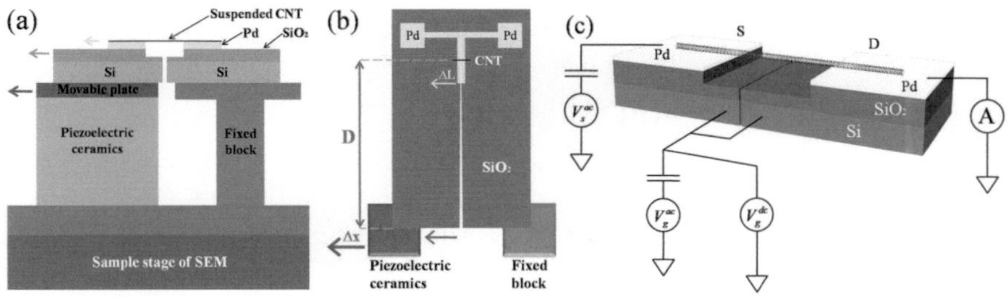


**(a)** Schematic of a piezoceramic-based setup used to induce axial strain in a CNT by displacing one electrode. The displacement is achieved via expansion of the piezoceramic block and measured in SEM. Reprinted with permission from [70]. ©2014 American Chemical Society.

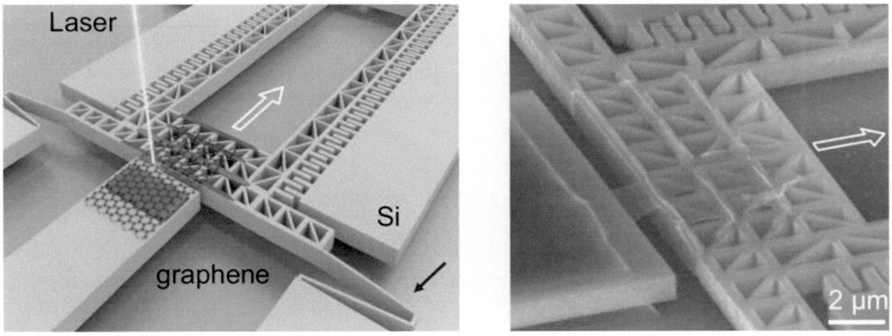


**(b)** Graphene strain tuning using a MEMS-based electrostatic comb actuator. Left: schematic MEMS. Right: SEM image highlighting the suspended comb structure and strained graphene and PMMA top clamping. This structure was not used for mechanical resonance experiments. Reprinted with permission from [71]. ©2018 American Chemical Society.

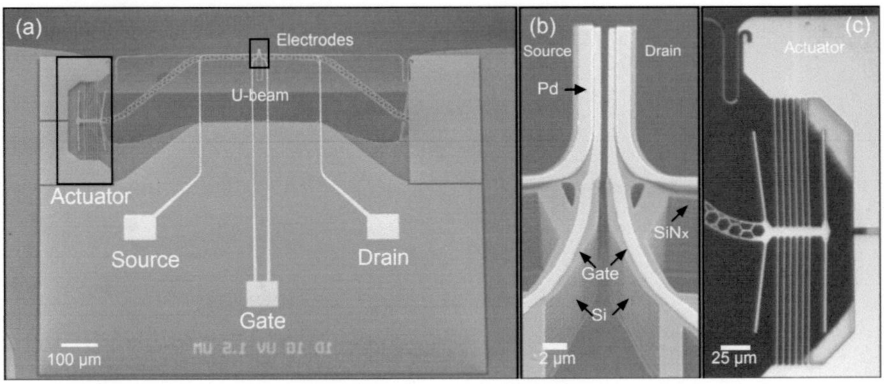


**(c)** Previous axial tuning approach developed by the Micro- and Nanosystems group, showing a suspended CNT device strained via a thermal actuator (c). The electrodes are seen in (b). This approach is further developed in this thesis. Open Access figure from [66] published under open access under CC BY licence (Creative Commons Attribution 4.0 International license).

**Figure 2.6:** Overview of different axial tuning mechanisms. The images illustrate various approaches for inducing strain in suspended nanomaterials such as CNTs and graphene. (a) employs piezoceramic actuation for electrode displacement; (b) uses electrostatic MEMS comb actuators to strain graphene; and (c) shows a thermal actuator setup.

excellent on-chip integration, precise control over strain, and low power consumption. However, drawbacks include the potential for electrostatic stiction or dielectric breakdown at higher voltages.

A second MEMS based approach to apply axial strain on resonating CNTs was presented in [72, 73] and further developed in [66, 74] (as shown in Fig. 2.6c). It consists of one thermal actuator connected to the end of one CNT electrode. The displacement is controlled via current which heats up the thermal actuators and displaces them via joule heating expansion.

Muoth et al. [72] used a similar platform for CNT strain experiments without resonance and achieved a displacement of 0.38 µm with 0.9 V applied actuator bias. They were able to strain the CNT up to 5 % and achieved 112 nm at 1 V with a maximum characterized range of up to 300 nm [74]. While Muoth et al. [72] were working towards static strain sensitivity, Truax et al. [73] used their setup to remove initial slack in the CNT of the resonator. Their measured eigenfrequencies of the first CNT mode tuned from 4 to 15 MHz and the second mode from 10 to 50 MHz.

This work builds upon the previous presented work on CNT MEMS straining platforms with thermal actuators and further improves design and functionality. A more detailed explanation of thermal actuators is given in chapter 4.

### 2.2.3 Overview of Achieved Tuning Ranges

A summary of achieved tuning ranges in the field of CNT resonators is given in table 2.1. Equation 2.1 is used with typical CNT values to visualize expected eigenfrequencies for different suspended lengths and strains in Figure 2.7. Selected tuning ranges presented in table 2.1 are highlighted as well. The achieved tuning ranges from literature were extracted from the available information in the different publications. The full tuning ranges were partially only available from figures, allowing for minor inaccuracies. The extracted tuning ranges were then plotted at the reported suspended CNT length of each publication.

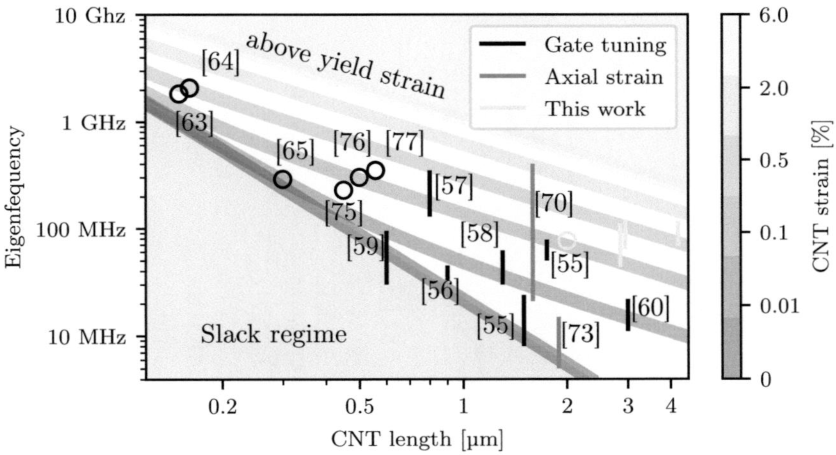


**Figure 2.7:** Overview of tuning mechanisms and performance in CNT resonators across the literature. The plot shows the reported eigenfrequencies as a function of the reported CNT suspended length for the two tuning methods. Vertical bars indicate the tuning range achieved by gate tuning (black), axial strain (red), and this work (cyan). Shaded background regions indicate approximate theoretical contours of constant CNT strain (based on Equation 2.1). The slack regime and above yield strain regions are marked. This comparison highlights both the tunability and limitations of different actuation approaches in CNT-based MEMS.

As expected from equation 2.1, shorter CNTs achieve higher eigenfrequencies. Many presented CNT resonators exhibit slack. It can also be seen that the presented gate bias tuning approaches either can not remove the slack fully or can barely strain above 0.1 %, whereas axially strained CNTs almost strain up to 2 % (with the exemption of Truax et al. [73] who were focused on slack removal).

**Table 2.1:** Extracted CNT resonator parameters from literature.

| | Authors | Year | $f_0$ | $Q$-factor | Tuning range | CNT length | $T$ | $p$ |
|---|---|---|---|---|---|---|---|---|
| Room temperature | Jensen [77] | 2006 | 300 MHz | 1000 | 100 MHz | 558 nm | RT | $< 10^{-7}$ torr |
| | Poot [59] | 2007 | 30 MHz | - | 35 MHz | 600 nm | RT | - |
| | Lassagne [56] | 2008 | 33 MHz | $< 200$ | 12 MHz | 900 nm | 300 K | $< 10^{-6}$ mbar |
| | Witkamp [58] | 2006 | 30 MHz | $< 160$ | 32 MHz | 1.3 µm | RT | $< 10^{-5}$ mbar |
| | Ning [70] | 2014 | 21 MHz | 312 | 410 MHz | 1.59 µm | 298 K | $< 10^{-5}$ torr |
| | Aykol [54] | 2014 | 10 MHz | $< 10^3$ | 27 MHz | 1.6 µm | 298 K | $< 10^{-8}$ torr |
| | Sazonova [55] | 2004 | 55 MHz | 80 | 28 MHz | 1.75 µm | 298 K | $< 10^{-5}$ torr |
| | Eichler [61] | 2013 | 52 MHz | $< 70$ | 16 MHz | 1.8 µm | 300 K | - |
| | Truax [73] | 2014 | 10 MHz | $< 100$ | 50 MHz | 1.89 µm | 300 K | vacuum |
| | Wei [78] | 2009 | 10 MHz | $< 1200$ | 25 MHz | 6.4 µm | RT | $< 10^{-5}$ torr |
| | 2 devices | This work | 74 MHz | $< 200$ | 4 MHz | 2 µm | RT | $< 10^{-3}$ mbar |
| | chap. 5 [79, 80] | | 37.4 MHz | $< 200$ | 3.6 MHz | 2 µm | RT | $< 10^{-3}$ mbar |
| | 3 devices | | 68.3 MHz | - | 49 MHz | 4.2 µm | RT | $< 10^{-3}$ mbar |
| | see chapter 6 | | 43.3 MHz | $< 300$ | 66.8 MHz | 2.9 µm | RT | $< 10^{-3}$ mbar |
| | [81] | | 63.5 MHz | $< 300$ | 57.4 MHz | 2.9 µm | RT | $< 10^{-3}$ mbar |
| Cryogenic temperature | Chaste [63] | 2012 | 1.84 GHz | - | 60 MHz | 150 nm | 6 K | $< 3 \times 10^{-11}$ mbar |
| | Wang [64] | 2018 | 2.08 GHz | 8500 | 120 MHz | 161 nm | 30 mK | $< 10^{-6}$ torr |
| | Häkkinen [65] | 2015 | 280 MHz | $1.3 \times 10^4$ | 18 MHz | 300 nm | 4 K | - |
| | Chiu [75] | 2008 | 225 MHz | 200 | 10 MHz | 450 nm | 6 K | high vacuum |
| | Gomez [82] | 2012 | 175 MHz | 15000 | 3 MHz | 600 nm | 4.2 K | - |
| | Hüttel [57] | 2009 | 120 MHz | 140670 | 240 MHz | 800 nm | 20 mK | - |
| | Eichler [83] | 2011 | 254 MHz | $> 10^4$ | 1.2 MHz | 840 nm | 90 mK | high vacuum |
| | Ning [84] | 2016 | 16.6 MHz | $< 150$ | 77.8 MHz | 1.6 µm | cryogenic | high vacuum |
| | Moser [85] | 2014 | 25 MHz | $5 \times 10^6$ | 45 MHz | 1.8 µm | 30 mK | - |

## 2.3 Current Research Objectives and Challenges

While their potential tuning range is very promising, CNT resonators also face certain weaknesses and challenges. This section discusses continuum modelling of CNTs, the quality factor, and CNT-electrode contact to highlight these obstacles and identify remaining research gaps.

### 2.3.1 Modelling of the CNT as Continuum

CNTs can often be treated as continuum structures owing to their extremely large aspect ratio. At length scales much greater than atomic dimensions, the discrete nature of a CNT becomes less significant, and its mechanical behaviour approaches that of a continuous medium. In continuum models, a CNT is thus represented not by discrete atoms but as an elastic continuum with effective material properties (like represented in Eqn. 2.1).

For precise modelling, information of the CNT length $l$, cross-sectional area $A_c$, moment of inertia $I_y$, radius $r$, Young's modulus $E$, and mass density $\mu$ is necessary. While information about the length and the radius can directly be measured via e.g. SEM or AFM, the cross-sectional area and the moment of inertia depend directly on the chosen continuum model. Previous work modelled the CNT either as a cylindrical, but filled beam (rod) or, as the name would suggest, as a tube. The approximation of the complex CNT structure into a continuum tube model is visualized in Figure 2.8. If a tube model is chosen, the question about its respective wall thickness rises.

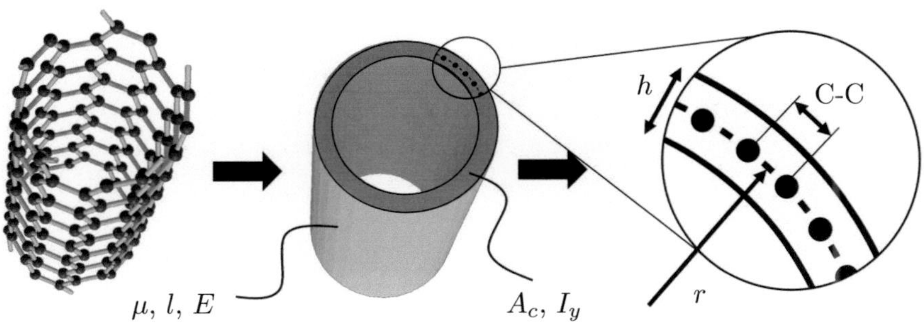


**Figure 2.8:** Visualization of the continuum model approximation. The CNT has been rendered using Blender CNT.

Table 2.2 displays a non exhaustive list of successfully applied models for describing mechanical behaviour in experiments. Many tube models use 340 pm as wall thickness for the CNT [55, 78, 86–88], a value that originates from the inter-layer distance of multi walled CNTs (MWCNTs) [89] or the thickness of a graphene sheet [90]. Only few references in literature have used thinner CNT tube models, partially without further justification [73,83,91]. Joshi et al. [91] justify their wall thickness of 61.7 pm, because they expect that the effective wall thickness of a continuous nanotube must be smaller than the diameter of a carbon atom ($< 183$ pm). They explain that stresses in the CNT are transmitted through a limited number of atoms, while in the continuum mechanics model the same stresses are transmitted through the wall area.

To put these sizes into context, the CNT suspended lengths range between 150 nm and 5 µm (Tab. 2.1), diameter lie usually between 1.5 nm and 2.5 nm [92], and the $SP^2$ C-C bond length between two neighbouring atoms is 142 pm [93]. The carbon atom has a diameter of around 183 pm and the carbon nucleus has a diameter of 5.4 fm ($5.4 \times 10^{-15}$ m).

The used Young's moduli in Table 2.2 vary between 0.1 TPa and 4.8 TPa, a quite large range. Experiments to determine the effective Young's modulus of a SWCNT include the observation of their free-standing vibrations in a TEM [94], using an AFM tip for controlled deflection of a doubly clamped CNT [50,95], using axial strain and measure displacement and force [96], or using suspended SWCNTs over a slit subjecting them to a calibrated transverse force and measuring the resulting transverse displacement [97]. To extract the Young's modulus from the data, all studies need to make assumptions about the cross sectional area of the SWCNT. All of the mentioned assume a wall thickness of the SWCNT of 340 pm. Measurement results yielded values usually between 1 TPa and 1.25 TPa, which is therefore only valid in combination with the wall thickness used to derive the Young's modulus. Over time, the CNT resonator community agreed to use 1 TPa as a standard value, which is the same as for graphene [90,98].

To understand how the fitting of a continuum model onto data can yield such different results, the CNT can be modelled in terms of bending rigidity $EI$ and longitudinal rigidity $EA$, that both depend on the cross-sectional properties of the continuum model (see also Section 3.5) [109]:

Cylinder model

$$EA = E\pi r^2 \tag{2.2}$$

$$EI = E\pi/4r^4 \tag{2.3}$$

Tube model

$$EA = 2\pi r E h \tag{2.4}$$

$$EI \approx \pi r^3 E h \tag{2.5}$$

For the tube model, it can be seen, that for each individual stiffness $E$ and $h$ can be traded-off, i.e. a smaller $E$ and a larger $h$ can yield the same stiffness as a larger $E$ and a smaller $h$. Joshi et al. [91] modelled the CNT as a pure beam, while Truax et al. [73] and Eichler et al. [83] modelled the CNT as a pure string, using only one of the rigidities respectively. This could explain the discrepancy. However, also when modelling both rigidities, the applied

**Table 2.2:** Overview of selected continuum models from literature. The table summarized used parameters for modelling and compares bending stiffness $EI$ and axial rigidity $EA$ values used in various continuum descriptions of CNTs. Where not explicitly provided in the original references, $EI$ and $EA$ were calculated based on the reported parameters using the standard beam theory formulas (Eqn. 2.2-Eqn. 2.5). The radius of Device MV 01 carries a large uncertainty and is discussed in Chapter 6. Data extracted from combined strain experiments on devices MV 12 and MV 39 hints towards a reduced product of $E \cdot h$. Further details are provided in Section 6.3

| Model | $E$ TPa | $r$ nm | $h$ pm | $EI$ $10^{-24}$ N m$^2$ | $EA$ µN | $Eh$ N m$^{-1}$ | Reference |
|---|---|---|---|---|---|---|---|
| Cylindrical beam | 1.25 | 1.5 | | 4.97 | 8.84 | | [57] |
| | 1.25 | 1.4 | | 3.77 | 7.70 | | [58] |
| | 1.0 | 0.55 | | 0.07 | 0.95 | | [99] |
| | 1.0 | 0.9 | | 0.52 | 2.54 | | [100] |
| | 1.0 | 1.2 | | 1.63 | 4.52 | | [101] |
| Hollow tube | 1.0 | 1 | 340 | 1.07 | 2.14 | 340 | [55] |
| | 1.2 | 1 | 340 | 1.28 | 2.56 | 408 | [86] |
| | 1.0 | 0.5 | 340 | 0.13 | 1.07 | 340 | [87] |
| | 1.2 | 1 | 340 | 1.28 | 2.56 | 408 | [88] |
| | 1.0 | 2 | 160 | 4.02 | 2.01 | 160 | [83] |
| | 0.1 | 0.5 | 80 | 0.003 | 0.03 | 8 | [73] |
| | 4.8 | 0.4 | 61.7 | 0.06 | 0.74 | 296.2 | [91] |
| Simulation | 4.9 | 0.7 | 61.7 | | | 302.3 | [102] |
| | 5.5 | 0.5 | 66 | | | 363 | [103] |
| | 5.5 | 0.61 | 66 | | | 363 | [104] |
| | 5.1 | | 74 | | | 377.4 | [105] |
| | 4.7 | | 75 | | | 352.5 | [106] |
| | 6.85 | | 80 | | | 548 | [107] |
| | 3.9 | | 89 | | | 347.1 | [108] |
| This work | 1.0 | 1.1 | 340 | ? | ? | 340 | CGDS 21 |
| | 1.0 | 0.8 | 340 | ? | ? | 340 | SJDS 123 |
| | 1.0 | | 340 | | | | MV 01 |
| | 1.0 | 0.75 | 340 | | | | MV 12 |
| | 1.0 | 0.95 | 340 | | | | MV 39 |

strain and the pre-strain are unknown, especially as the applied strain by the gate also depends on $EA$ (see Section 3.3.1).

We can see, that independently all products $E \cdot h$ agree well over all different studies (with the exception of [73, 83, 107]) and lie between 300 and 400 $\mathrm{N\,m^{-1}}$. The importance of this product becomes visible in Equations 2.4 and 2.5.

SWCNTs have also been simulated in molecular dynamic simulations (Tab. 2.2). The simulations consistently result in wall thicknesses much thinner than 340 pm, usually under 100 pm [110]. The Young's moduli of these simulations are much larger than 1 TPa, which is a value that the experimental CNT resonator community commonly uses [55, 83, 99–101]. However, the product of Young's modulus and the wall thickness remains similar.

Accurate knowledge of the applied strain, together with resonance experiments, can help clarifying discrepancies between values obtained from molecular simulations and those measured in the resonator domain, as well as reconcile differences reported within the resonator community.

### 2.3.2 Quality Factor

The quality factor $Q$ quantifies how efficiently a resonator stores energy. A high $Q$ indicates low energy loss over time and corresponds to a narrow resonance peak, while a low $Q$ describes broader peaks and greater energy dissipation. It is defined as the ratio of energy stored in the system to the energy lost per oscillation cycle:

$$Q = 2\pi \frac{Energy_{\mathrm{stored}}}{Energy_{\mathrm{lost\ per\ cycle}}} \tag{2.6}$$

As shown in Table 2.1, CNT resonators typically reach quality factors in the hundreds at room temperature and under vacuum. Molecular simulations predict a room temperature limit of 2,000–3,000, attributed to bond-breaking and reforming between carbon atoms [111,112]. At cryogenic temperatures, quality factors as high as 6.8 million have been demonstrated at 70 mK [85,113]. However, the measured $Q$ values remain far below theoretical limits due to multiple dissipation mechanisms, generally classified as intrinsic or extrinsic.

Intrinsic losses originate from the material itself and may be mechanical, electrical, or thermal in nature, occurring either on the surface or within the bulk. These are subdivided into frictional and fundamental losses [53]. For nanoscale resonators like CNTs, surface friction is dominant due to their high surface-to-volume ratio [114]. Energy loss mechanisms include adsorbates moving on the surface or are due to defects and roughness. In CNT resonators the breaking and reformation of covalent bonds between the carbon atoms is the highest energy loss mechanism [115]. Other friction losses for CNT resonators come from irreversible dislocation of defects in their crystal structure. This class of energy loss can be reduced by optimizing the quality of the resonating material.

Fundamental losses are unavoidable. The first, thermoelastic damping, arises from localized heating during oscillations due to thermal expansion. This effect is negligible in CNTs due to their high resonance frequency that oscillates too fast for heat transfer to occur [116]. The second, Akhiezer damping (phonon-phonon interaction loss), occurs when strain from vibration perturbs phonon equilibrium, leading to mechanical energy dissipation via phonon-phonon scattering.

Extrinsic losses involve interactions with the environment or clamping structure. Medium interaction loss, or gas damping [117], is working in a ballistic or a fluidic regime, determined by the Knudsen number, which relates the mean free path of the gas with the physical length scale of the resonator. In Table 2.1, all measurements were conducted in vacuum (typically below $1 \times 10^{-3}$ mbar), minimizing damping (ballistic regime). Fluidic damping behaviour has been experimentally investigated at higher pressures (1 mbar) by Kumar et al. [66].

Anchor loss, the dominant extrinsic dissipation mechanism in CNT resonators, refers to energy leakage through the clamping points. While perfect clamping is unachievable, energy loss can be reduced by modifying substrate geometry to alter phonon coupling, or by employing phononic crystals with bandgaps around the resonance frequency - though not yet applied to CNTs [118]. Anchor losses remain a primary bottleneck [119].

Approaching the theoretical $Q$ limit is crucial for developing high-performance devices. Current efforts focus on reducing both intrinsic and extrinsic losses, particularly anchor and interaction losses.

According to Equation 2.6, $Q$ can also be increased by enhancing stored energy, for example, through elastic strain. This approach, also known as damping or *dissipation dilution*, has been widely explored in various systems, including silicon nitride strings [120]. In CNTs, Ning et al. demonstrated an increase in $Q$ from approximately 75 to 300 under 2 % applied strain [70].

### 2.3.3 CNT Contacts to Electrodes

The contacts between the CNT and the electrode play a crucial role for device performance, because they define the contact resistance, a dissipation path of energy, and the mechanical contact strength. For the eigenfrequency tuning of CNT resonators, the mechanical strength is essential, because the contacts need to withstand the built up tension in the CNT with applied strain.

Two contact types resulting from the fabrication approach were presented in Section 2.2.1. The bottom-up approach uses electron beam lithography (e-beam) to make electrical contacts on top of the CNT after localizing it. This results in a CNT that is clamped from top by a metal such as gold and from bottom by the substrate as seen in Figure 2.5a and Figure 2.5b. [55, 58].

The second clamping type results from dry-transferred CNTs and is known as "adhesively clamped" where Van der Waals forces between the CNT and the electrodes are used to hold the CNT in place [60, 79, 80]. After this "bottom clamped" state, further treatment is possible. Kumar et al. [119] investigated selective platinum (Pt) atomic layer deposition (ALD) methods for clamping, while other researchers have explored focused ion beam (FIB) deposition of Pt. Ning et al. [70] employed e-beam induced carbon deposition to anchor the CNT to the substrate, achieving the highest internal strain observed to date in CNT resonators.

Effective clamping is not only crucial for achieving high strain levels and large tuning ranges, but it also plays a significant role in maintaining a high quality factor, as most energy dissipation in CNT resonators is attributed to anchor losses. Kumar et al. [119] demonstrated that improved top clamping increased the quality factor ($Q$-factor) of CNT resonators by a factor of 1.5, reaching values up to 220. Ning et al.'s method of top clamping, while effective at low strain levels, yielded $Q$-factors below 150, although values near 300 were achieved with dissipation dilution.

The total electrical resistance of a CNT resonator device consists of the CNT channel resistance, the CNT quantum resistance and the contact resistances between the CNT and the electrodes (contacts). The total CNT resistance is defined by the quantum resistance of around $6.45\,\mathrm{k\Omega}$ [121] and the CNT channel resistance, which is chirality dependent, reaches tens of $\mathrm{k\Omega}$. The main resistance contribution to the total device resistance is due to the contact resistances, which can reach up to $\mathrm{M\Omega}$. Low contact resistances are desirable for improved signal-to-noise-ratio. Low signal-to-noise ratio can prevent the measurement of resonance, as the signal is overshadowed by the noise.

Contact resistances can be reduced with several different treatment methods as proposed by Jung et al. [92]. Pre-transfer treatment can consist of cleaning the electrodes to remove native oxide from the electrodes by Argon-ion etching (e.g. 25 W for 1 min) shortly before the CNT dry transfer.

To further reduce the contact resistance, the device can be annealed in vacuum at 250 °C or in a nitrogen or argon environment [122]. The combination of both treatments for dry transferred CNT devices have yielded contact resistances below 200 kΩ [123].

Good mechanical contacts ensure long-term stability and performance of CNT resonators. The mechanical contacts can be evaluated by estimating the highest achieved tension in the CNT. Reported tension values from literature have to be treated with caution, as they might be based on different CNT wall thickness, different continuum models, or different assumed Young's moduli (as introduced in Section 2.3.1).

The highest tension in CNT resonators was reported by Ning et al. [70]: they reached an internal tension $T_i$ of 7.24 nN by axially applied strain levels $\epsilon_A$ of around 2 %. Kumar et al. [119] reported an internal tension $T_i$ of 0.64 nN (using $T_i = \sigma \cdot A$ and reported a CNT internal stress of $2.5 \times 10^8\,\mathrm{N\,m^{-2}}$). Gate bias induced tensions have also been studied by Kumar et al., where the application of a gate bias led to a tension of up to 2.6 nN, which triggered a slippage event that caused a reduction in the eigenfrequency of the CNT resonator due to stress relaxation [101]. This slippage effect highlights the need for improved clamping mechanisms to enhance stability. Similar slippage events have been observed in this work and will be discussed in Chapter 6.

In this chapter, an overview of different MEMS resonator architectures was given. Further, CNTs were identified as the most promising material towards achieving large tuning ranges. Both gate bias tuning and axial strain tuning were introduced to tune the eigenfrequencies of CNTs and achieved tuning ranges in literature were presented. The split in the CNT resonator community between rod and tube modelling of the CNT was highlighted. Substantial differences between the experimental results and molecular dynamic simulations were mentioned, highlighting the need for further research.

## 2.4 Thesis Motivation and Outline

CNT resonators exhibit an exceptionally large potential for eigenfrequency tuning, surpassing what is achievable with most conventional resonator types. Despite this promise, fully harnessing their tunability has proven difficult, in large part because applying axial strain through a MEMS poses design and fabrication challenges. While gate-bias-induced strain has been widely used in CNT resonator studies, only a single demonstration of an axially tunable CNT resonator exists, relying on a bulky actuator that only permits measurements inside a scanning electron microscope. Further, significant inconsistencies persist in how CNTs are modelled in the literature: some researchers treat them as solid rods, whereas others describe them as hollow tubes. There is also a disagreement on the appropriate wall thickness to use. This issue was highlighted by divergent results in molecular simulations versus experimental resonator work.

This thesis tackles these challenges by creating a MEMS-based straining platform that can apply pure axial strain to a CNT and also combine it with gate-bias-induced strain. By reassessing existing continuum models, carefully defining strain levels, and integrating both straining methods into a single device, it aims to reduce uncertainty in key CNT parameters while also aiming to extend the tuning range. The thesis is organized as follows:

- **Chapter 3** presents a comprehensive overview of modelling CNT resonators as doubly clamped mechanical continua, including solutions to nonlinear beam equations. It shows how different forms of strain - gate bias induced and axial applied - modify the resonator's behaviour and demonstrates a powerful numerical approach to handle these nonlinearities.

- **Chapter 4** introduces the MEMS straining platform and discusses its design considerations, fabrication steps, and structural features that allow accurate axial displacement of the CNT's clamping ends. The chapter also explains the characterization process using scanning electron microscopy and supports the observations with finite element method (FEM) simulations that capture the thermomechanical behaviour of the chevron-style actuators.

- **Chapter 5** explores the methods for detecting and analysing motional currents in CNT resonators under gate-bias strain. It applies the nonlinear modelling techniques from Chapter 3 to experimental data, thereby validating the theoretical approaches and showcasing how gate-bias strain affects the resonator's electrical readout.

- **Chapter 6** reports on CNT resonators integrated into the MEMS platform to achieve precise axial strain. It presents pure axial strain measurements that extract unknown parameters (such as the pre-strain) and follows with experiments that combine both axial and gate-bias strain. The chapter addresses how these results compare with previously assumed mass-loading values and discusses the implications for modelling the CNT's longitudinal stiffness and overall structural dynamics.

- **Chapter 7** concludes the thesis, summarizing the major findings, highlighting the improved understanding of CNT strain engineering and modelling, and suggesting future research directions to further expand the performance and scalability of tunable CNT resonators.

# 3 Mechanics of Doubly Clamped Resonators

In this chapter, the theory regarding the mechanical behaviour of CNT resonators is presented. First, a general derivation of the physics is shown, valid for any kind of beam, and then this is tailored step-by-step to fit the application of CNTs. The influence of tension on the general behaviour of resonance is investigated and put into context with the beam-string transition. Then, boundary condition influences are included in the general differential equation of motion to describe CNT resonators specifically.

After the general mechanic models have been applied, the Duffing equation, a nonlinear differential equation used for lumped mechanical oscillators, is discussed. It is shown, that the general equation of motion can be translated into this well-studied equation. This requires deeper analysis of the beam's tension, which explains and separates the two tuning approaches *gate bias tuning* and *axial strain tuning* presented in Section 2.2. Further, the actuation method and actuation force are introduced.

Then the solution of the Duffing equation is presented. It describes the motion of the resonator and is hence required for modelling its behaviour. In this work, harmonic balancing (HB) is used to approximate the solution. The influences of gate bias tuning and axial strain tuning are modelled and discussed. Lastly, continuum modelling of CNTs is presented.

## 3.1 Flexural Vibrations in Beams

The objective of studying mechanical resonators is to determine their *eigenfrequencies* and *eigenmodes* to characterize their natural vibrational behaviour. Eigenfrequencies refer to the natural frequency of oscillation, while the eigenmode describes the spatial pattern of motion. They result from Newton's third law applied on an infinitesimal small segment of the resonating structure. This results in a differential equation describing how inertia, elastic properties, and external forces interact. Its solution yields the eigenfrequencies and describes vibrational profiles $z(x,t)$, where $x$ denotes the position and $t$ the time. This derivation has been shown in

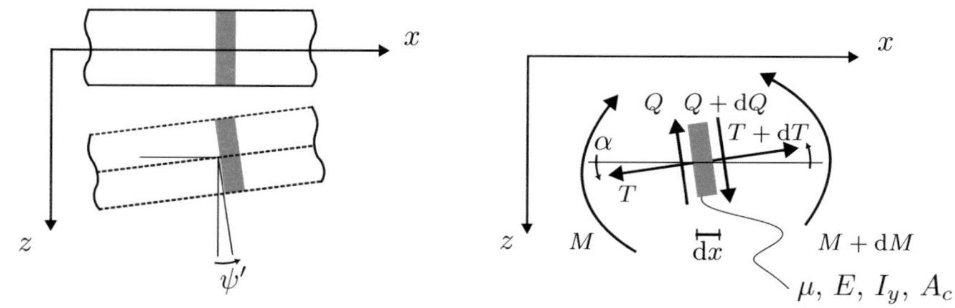


**(a)** Deflected shape of a beam section. **(b)** Free-body diagram of a beam segment.

**Figure 3.1:** Deflected shape and free-body diagram of a beam section with internal forces and moments.

literature [53, 66]. Commonly, the derivation is expressed in terms of axial stresses $\sigma$ and volumetric mass density $\rho$, but because the CNT is one-dimensional, the linear mass density $\mu = \rho A_c$ and tension $T = \sigma A_c$ are used in this work.

Figure 3.1a shows a sketched beam in deflected and undeformed position with a highlighted section of an infinitesimal small section. A free-body-diagram is shown in Figure 3.1b with the acting tension $T$, shear force $Q$, and bending moment $M$ on the section with linear density $\mu$, Young's modulus $E$, moment of inertia $I_y$, and cross-sectional area $A_c$. Applying Newton's third law on the infinitesimal small section yields:

$$\sum F_x = 0 \qquad : \quad dT = 0 \tag{3.1}$$

$$\sum F_z = 0 \qquad : \quad \mathrm{d}Q = -\mu \frac{\partial^2 z}{\partial t^2}\,\mathrm{d}x \tag{3.2}$$

$$\sum M_y = 0 \qquad : \quad Q = \frac{\mathrm{d}M}{\mathrm{d}x} - Tz'(x) \tag{3.3}$$

From the force balance in beam direction (Eqn. 3.1), it can be seen that the tension force does not change along the beam. Using the relation between bending moment $M$, curvature $\frac{\partial^2 z}{\partial x^2}$ and deflection $z$, the derivative of the bending moment $M$ can be expressed as:

$$\frac{\mathrm{d}M}{\mathrm{d}x} = EI_y \frac{\partial^3 z}{\partial x^3} \tag{3.4}$$

Combining this equation with the force and moment equilibria (Eqn. 3.1-3.3) yields the differential equation of motion:

$$\mu\frac{\partial^2 z(x,t)}{\partial t^2} + EI_y\frac{\partial^4 z(x,t)}{\partial x^4} - T\frac{\partial^2 z(x,t)}{\partial x^2} = 0 \tag{3.5}$$

In Figure 3.1, rotational inertia and shear deformation are neglected, because CNTs, as 1-D materials, are by definition slender ($l/d > 1000$) and therefore its bending dynamics can be described by the Euler Bernoulli beam theory instead of the Timoshenko beam theory. Due to the high length-to-diameter ratio of the CNT and the lack of radial confinement, the stress state can be considered uni-axial. This justifies the use of the true Young's modulus $E$ , rather than the plane stress equivalent $E/(1-\nu^2)$.

This differential equation is solved by using the ansatz of "separating the variables". The solution consists of a spacial and a time-dependent term and is a superposition of all eigenmodes, expressed as

$$z(x,t) = \sum_{n=1}^{\infty} \phi_n(x) \cdot z_{0,n}\cos(\omega_n t) \tag{3.6}$$

where $n$ is the mode, $\phi_n$ is the mode shape of the $n^{th}$ eigenmode, $\omega_n$ is the eigenfrequency of the $n^{th}$ eigenmode, and $z_{0,n}$ is the maximum amplitude of oscillation in that mode. The modeshape $\phi_n$ is normalized to have a maximum of 1. The general solution for the mode shape has the form:

$$\phi_n(x) = \underbrace{a_n\cos(\beta_n x) + b_n\sin(\beta_n x)}_{=S \text{ (standing wave)}} + \underbrace{c_n\cosh(\beta_n x) + d_n\sinh(\beta_n x)}_{=C \text{ (clamping)}} \tag{3.7}$$

where $\beta_n$ is the wave number. The trigonometric functions embody standing waves in the beam and are denoted by $S$, the hyperbolic functions include the clamping conditions into the solutions and are denoted as $C$. Depending on the clamping conditions, the solution looks differently as shown in Section 3.2 and will result in different $a_n, b_n, c_n, d_n$.

Using the general solution of the mode shape function $\phi_n(x)$ (Eqn. 3.7) and the ansatz of the separation of variables (Eqn. 3.6) for the differential equation of motion (Eqn. 3.5) yields the *dispersion relation*:

$$-\omega^2\mu(S+C) + EI_y\beta_n^4(S+C) - T\beta_n^2(-S+C) = 0 \tag{3.8}$$

Separating the contribution of the standing wave and the clamping influence gives:

$$S\left(EI_y\beta_n^4 + T\beta_n^2 - \omega^2\mu\right) + C\left(EI_y\beta_n^4 - T\beta_n^2 - \omega^2\mu\right) = 0 \tag{3.9}$$

This equation is only true if both summands are zero. At this point, for better overview, the wave numbers are renamed according to their influence. The wave number $\beta_n$ in the term of the standing wave is denoted as $\beta_s$, the wave number in from the clamping contribution is renamed $\beta_c$. The *dispersion relation* (Eqn. 3.9) is written as system of equations:

$$EI_y\beta_s^4 + T\beta_s^2 - \omega^2\mu = 0 \tag{3.10}$$
$$EI_y\beta_c^4 - T\beta_c^2 - \omega^2\mu = 0 \tag{3.11}$$

These quadratic equations have both positive and negative solutions. Here, only the positive solutions are considered. The wave numbers are as follows:

$$\beta_s = \sqrt{-\frac{T}{2EI_y} \pm \sqrt{\left(\frac{T}{2EI_y}\right)^2 + \frac{\mu\omega^2}{EI_y}}} \tag{3.12}$$

$$\beta_c = \sqrt{\frac{T}{2EI_y} \pm \sqrt{\left(\frac{T}{2EI_y}\right)^2 + \frac{\mu\omega^2}{EI_y}}} \tag{3.13}$$

At this point, no boundary conditions have been imposed, so the wave numbers are general and depend on the beams geometry ($I_y$), material properties ($\mu, E$) and the external tension $T$. The tension influence is now discussed.

Equation 3.12 can be rearranged to express:

$$\omega_0 = \sqrt{\frac{EI_y}{\mu}\beta_s^4 + \frac{T}{\mu}\beta_s^2} = \beta_s^2 \cdot \sqrt{\frac{EI_y}{\mu}} \cdot \sqrt{1 + \frac{T}{EI_y\beta_s^2}} \tag{3.14}$$

$$f_0 = \frac{1}{2\pi}\sqrt{\frac{EI_y}{\mu}\beta_s^4 + \frac{T}{\mu}\beta_s^2} = \frac{1}{2\pi}\sqrt{\underbrace{\frac{EI_y}{\mu}\beta_s^4}_{\text{beam}} + \underbrace{\frac{\epsilon \cdot EA_c}{\mu}\beta_s^2}_{\text{string}}} \tag{3.15}$$

As Equation 3.12 yields two solutions, one term corresponds to the bending-dominated wave number, which approaches the classical beam eigenvalue when $T$ is small, and the second corresponds to the tension-dominated wave number when $T$ is large.

The tension $T$ dependent expression 3.15 allows the analysis of the extreme cases of high or low tension. If the tension is small

$$T << EI_y\beta_s^2, \tag{3.16}$$

Equation 3.14 simplifies to

$$\omega = \beta_s^2 \cdot \sqrt{\frac{EI_y}{\mu}}, \tag{3.17}$$

which is the eigenfrequency of a beam with flexural rigidity. If the tension is high

$$T >> EI_y\beta_s^2 \tag{3.18}$$

the Equation 3.14 can be simplified to

$$\omega = \beta_s \cdot \sqrt{\frac{T}{\mu}}, \tag{3.19}$$

which describes the eigenfrequency of a string. The more general form of Equation 3.12 is not just able to represent either beam or string behaviour, but also the tension dependent transitional regime in between both.

## 3.2 Boundary Conditions

The wavenumbers $\beta_s, \beta_c$ and eigenfrequencies $\omega_n$ can be computed for defined boundary conditions. These vary depending on the clamping. Examples are fixed-free (cantilever), pinned-pinned (string), fixed-fixed (beam) as presented below:

Cantilever:

$$\text{fixed at } x = 0: \quad \phi(0) = 0, \quad \phi'(0) = 0, \tag{3.20}$$

$$\text{free at } x = l: \quad \phi''(l) = 0, \quad \phi'''(l) = 0, \tag{3.21}$$

String:

$$\text{pinned at } x = 0: \quad \phi(0) = 0, \quad \phi''(0) = 0, \tag{3.22}$$

$$\text{pinned at } x = l: \quad \phi(l) = 0, \quad \phi''(l) = 0, \tag{3.23}$$

Beam:

$$\text{fixed at } x = 0: \quad \phi(0) = 0, \quad \phi'(0) = 0, \tag{3.24}$$

$$\text{fixed at } x = l: \quad \phi(l) = 0, \quad \phi'(l) = 0 \tag{3.25}$$

$\phi = 0$ means no deflection, $\phi' = 0$ no slope, $\phi'' = 0$ no momentum, and $\phi''' = 0$ no shear force. Inserting the boundary conditions of the beam configuration for the CNT (Eqn. 3.24 and 3.25) into the general solution for the modeshape (Eqn. 3.7) gives:

$$\phi(0) = 0: \quad a_n + c_n = 0 \tag{3.26}$$

$$\phi'(0) = 0: \quad b_n\beta_s + d_n\beta_c = 0 \tag{3.27}$$

$$\phi(l) = 0: a_n \cos(\beta_s l) + b_n \sin(\beta_s l) + c_n \cosh(\beta_c l) + d_n \sinh(\beta_c l) = 0 \tag{3.28}$$

$$\phi'(l) = 0: \quad -a_n\beta_s \sin(\beta_s l) + b_n\beta_s \cos(\beta_s l) + c_n\beta_c \sinh(\beta_c l) + d_n\beta_c \cosh(\beta_c l) = 0 \tag{3.29}$$

The four Equations 3.26 - 3.29 can be represented with a boundary condition (BC) matrix as a system of linear equations:

$$\underbrace{\begin{bmatrix} 1 & 0 & 1 & 0 \\ \cos(\beta_s l) & \sin(\beta_s l) & \cosh(\beta_c l) & \sinh(\beta_c l) \\ 0 & \beta_s & 0 & \beta_c \\ -\beta_s \sin(\beta_s l) & \beta_s \cos(\beta_s l) & \beta_c \sinh(\beta_c l) & \beta_c \cosh(\beta_c l) \end{bmatrix}}_{\text{BC matrix}} \begin{bmatrix} a_n \\ b_n \\ c_n \\ d_n \end{bmatrix} = \begin{bmatrix} 0 \\ 0 \\ 0 \\ 0 \end{bmatrix} \tag{3.30}$$

This system only has a solution if the determinant of the BC matrix is zero. The frequency equation (characteristic equation) can be expressed as:

$$\underbrace{2\beta_s\beta_c\left(\cos(\beta_s l)\cosh(\beta_c l)-1\right)}_{f(T,\omega)}-\underbrace{\left(\beta_c^2-\beta_s^2\right)\sin(\beta_s l)\sinh(\beta_c l)}_{g(T,\omega)}=0 \quad (3.31)$$

$$f(T,\omega)-g(T,\omega)=0 \quad (3.32)$$

Solving the frequency equation for given tension $T$ as well as geometrical and material values provides the different eigenfrequencies $\omega_{0,n}$, which are required to compute the wavenumbers $\beta_s$ and $\beta_c$. The frequency equation is nonlinear with multiple solutions that correspond to the eigenfrequencies of different modes. To solve for the eigenfrequencies, a bisection method is used and implemented in python. Once the eigenfrequencies are obtained, the mode shape $\phi_n(x,\omega_{0,n},T)$ can be computed.

The boundary conditions are then applied onto the general solution (Eqn. 3.7) to obtain the mode shape for the clamped-clamped beam under tension. The two boundary conditions at $x=0$ (Eqn. 3.26 and Eqn. 3.27) reduce the general equation to:

$$\phi_n(x)=a_n\left(\cos(\beta_s x)-\cosh(\beta_c x)\right)+b_n\left(\sin(\beta_s x)+\frac{\beta_s}{\beta_c}\sinh(\beta_c x)\right) \quad (3.33)$$

The fourth boundary condition (Eqn. 3.29) is used to further simplify the mode shape, as it yields the ratio between the two coefficients $a_n$ and $b_n$:

$$\phi'(l)=0: \quad \frac{b_n}{a_n}=\frac{-\beta_s\sin(\beta_s l)-\beta_c\sinh(\beta_c l)}{\beta_s\cdot\left(\cos(\beta_s l)-\cosh(\beta_c l)\right)} \quad (3.34)$$

The mode shape for doubly clamped beams then becomes:

$$\phi_n=a_n\left(\cos(\beta_s x)-\cosh(\beta_c x)+\frac{\beta_s\sin(\beta_s l)+\beta_c\sinh(\beta_c l)}{\cos(\beta_s l)-\cosh(\beta_c l)}\left(\sin(\beta_s x)-\sinh(\beta_c x)\right)\right) \quad (3.35)$$

The mode shape $\phi_n$ is normalized with $a_n$ that changes with mode numbers. For $n=1,2,3$, $a_n$ is $0.6297, 0.6626, 0.7112$. For higher mode numbers, $a_n$ converges to $1/\sqrt{2}$ [53, 66].

Once the eigenfrequencies are known, the corresponding wave numbers can be calculated. For typical CNT parameters, the tension dependence of the

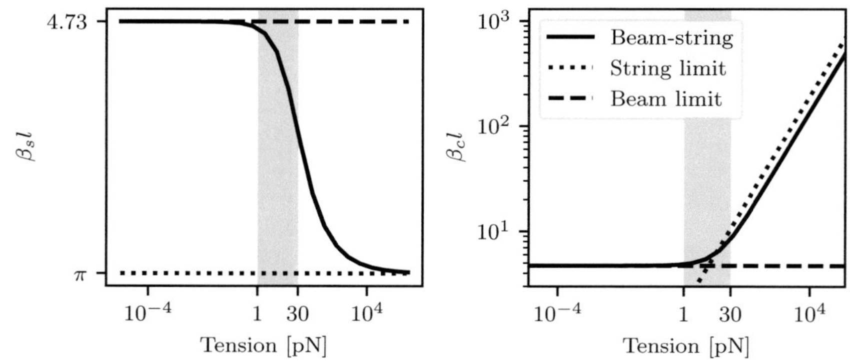


**Figure 3.2:** Scaled wave number $\beta_s$ by $l$ converges from 4.73 for low tensions to $\pi$ for higher tensions. $\beta_c$ diverges with higher tension. The beam and string limits are highlighted as well as a transition zone marked in light gray.

normalized wave numbers $\beta_s$ and $\beta_c$ are illustrated in Figure 3.2. The CNT properties used in these calculations are summarized in Table 3.1. As shown, the general expression converges to the pure beam behaviour at low tension, and to the pure string behaviour at high tension. Importantly, it also captures the transitional regime, where the resonator exhibits characteristics of both beam and string dynamics.

The eigenfrequencies of the exemplary CNT (Tab. 3.1) for different modes are shown in Figure 3.3. The influence of the tension on the eigenfrequency of the CNT becomes visible already around 1 pN or around $5 \times 10^{-5}$ % strain, highlighting the strain sensitivity of CNT resonators. The yield strain at 6 % is highlighted in red, indicating a maximum eigenfrequency of the example CNT around 1.3 GHz.

The influence of the tension on the boundary area is shown in Figure 3.4. As the tension is increased, the boundary region transitions from being dominated by bending to showing behaviour more characteristic of a taut string, altering both the mode shapes and the corresponding resonance conditions.

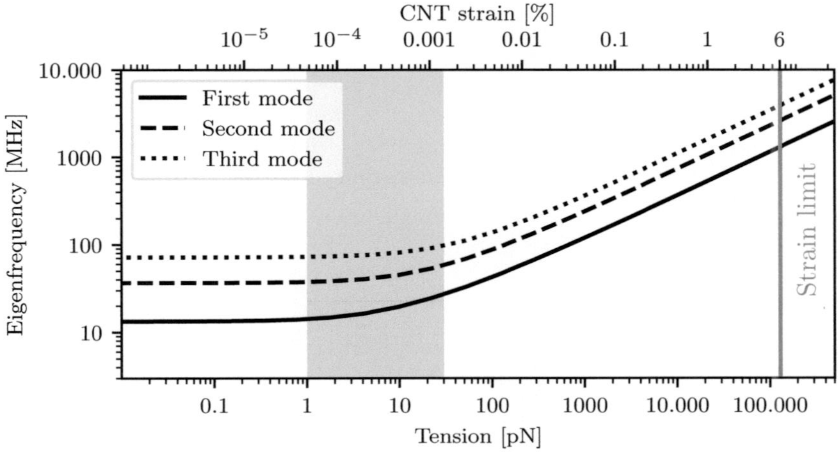


**Figure 3.3:** Eigenfrequencies of the first three eigenmodes of a CNT resonator with typical values dependent on tension or strain. The strain limit is highlighted.

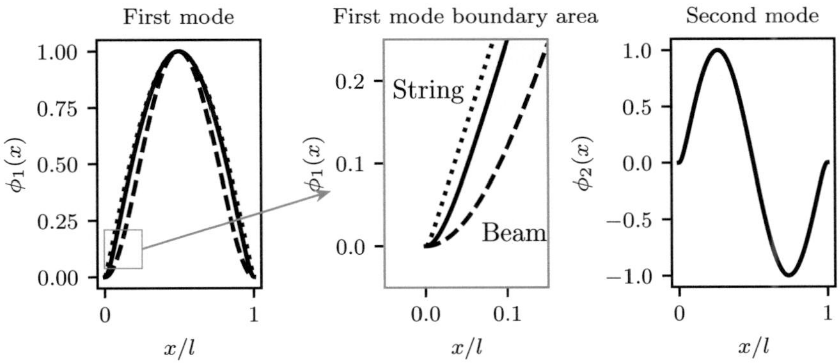


**Figure 3.4:** Mode shape of the first mode of a beam (dashed line), string (dotted line) and beam-string (solid line). Center: zoom in on the clamping region. Right: second mode of a doubly clamped resonator.

## 3.3 Modelling the Dynamics of CNT Resonators

After determining the tension dependent wavenumbers, the eigenfrequency and the mode shape of the resonator, it is necessary to solve the differential equation of motion (Eqn. 3.5) to describe the behaviour of the oscillating system. The response of the oscillating system is also dependent on external forces. Substituting the separation-of-variables ansatz (Eqn. 3.6) into Eqn. 3.5, and adding both a viscous-damping term $b\phi_n(x)\frac{\partial z(t)}{\partial t}$ and an external forcing term $f(x,t)$, yields:

$$\mu\phi_n(x)\frac{\partial^2 z(t)}{\partial t^2} + EI_y z(t)\frac{\partial^4 \phi_n(x)}{\partial x^4} - Tz(t)\frac{\partial^2 \phi_n(x)}{\partial x^2} + b\phi_n(x)\frac{\partial z(t)}{\partial t} = f(x,t) \tag{3.36}$$

Equation 3.36 (a partial differential equation) is transformed into a lumped parameter form (a set of ordinary differential equations) using Galerkin's method (see Appendix A.1) [53]. The method involves multiplying the Equation 3.36 by the modeshapes $\phi_n(x)$ and integrating over the domain from 0 until $l$. The equation is transformed using the method of integration by parts, yielding:

$$\begin{aligned}\mu\int_0^l \phi_n^2(x)dx\frac{\partial^2 z(t)}{\partial t^2} + b\int_0^l \phi_n^2(x)dx\frac{\partial z(t)}{\partial t} + EI_y z(t)\int_0^l \left(\frac{\partial^2 \phi_n(x)}{\partial x^2}\right)^2 dx \\ + Tz(t)\int_0^l \left(\frac{\partial \phi_n(x)}{\partial x}\right)^2 dx = \int_0^l f(x,t)\phi_n(x)dx\end{aligned} \tag{3.37}$$

Effective parameters are defined as:

$$m_{eff} = \mu \int_0^l \phi_n^2(x)dx \tag{3.38}$$

$$b_{eff} = b \int_0^l \phi_n^2(x)dx \tag{3.39}$$

$$k_{eff} = \underbrace{EI_y \int_0^l \left(\frac{\partial^2 \phi_n(x)}{\partial x^2}\right)^2 dx}_{\text{Beam}} + \underbrace{T \int_0^l \left(\frac{\partial \phi_n(x)}{\partial x}\right)^2 dx}_{\text{String}} \tag{3.40}$$

$$F_{eff} = \int_0^l f(x,t)\phi_n(x)dx \tag{3.41}$$

and are used to rewrite Equation 3.37 as:

$$m_{eff}\ddot{z}(t) + b_{eff}\dot{z}(t) + k_{eff}z(t) = F_{eff} \tag{3.42}$$

which resembles the differential equation of motion of an harmonic oscillator. The stiffness term $k_{eff}$ is further discussed in Section 3.3.1, the excitation force $F_{eff}$ in Section 3.3.2, and the effective damping rate $b_{eff}$ in Section 3.3.3.

The analysis of the first vibrational mode of a clamped-clamped beam under axial tension reveals that the behaviour of integrals over the mode shape is tension dependent and shows distinct trends depending on the derivative order involved. The integrals $\int_0^l \phi_n^2(x)dx$ and $\int_0^l \left(\frac{\partial \phi_n(x)}{\partial x}\right)^2 dx$, which correspond to the modal mass and slope (strain) energy respectively, both tend toward constant values as the axial tension increases. This reflects the fact that, although the mode shape becomes increasingly sinusoidal with rising tension, its overall spatial extent and the distribution of displacement remain bounded. As such, the energy contributions from displacement and slope do not grow indefinitely with tension, but instead stabilize once the beam behaviour transitions toward that of a taut string.

In contrast, the integral $\int_0^l \left(\frac{\partial^2 \phi_n(x)}{\partial x^2}\right)^2 dx$, representing the bending energy stored in the beam, increases monotonically with tension and does not exhibit bounded behaviour. While the central region of the mode shape flattens under high tension - resembling the string limit - the clamped boundary conditions require the displacement and slope to vanish at both ends of the beam. As a result, the curvature near the boundaries becomes increasingly localized and sharp in order to satisfy these constraints. This leads to elevated second derivatives in narrow regions, causing the integral over $\left(\frac{\partial^2 \phi_n(x)}{\partial x^2}\right)^2$ to grow significantly with tension. Therefore, unlike the other integrals, the bending energy does not approach a limiting value but continues to increase as tension rises, reflecting the localized geometric adaptation required at the beam ends.

### 3.3.1 Effective Stiffness of a Clamped-Clamped Beam under Tension

The effective stiffness $k_{eff}$ of the resonator (Eqn. 3.40) is composed of the bending stiffness $EI_y$, which represents the beam behaviour and a tension $T$ which can influence the behaviour towards more string-like behaviour, as previously shown. The tension is used to tune the eigenfrequency and the working principles are presented in this section.

The tension $T$, in the resonator potentially consists of three elements: the pre-tension $T_0$, that might be induced in the CNT during the fabrication, an axially applied tension $T_{ax}$, and a tension induced through a transversally applied force by a static gate bias $T_g$. As the CNT is oscillating, tension is also induced. This effect will become apparent when looking closer at $T_g$. The stiffness influence of applied tension $T_{ax}$ and $T_g$ and pre-tension is then:

$$k_{tension} = (T_0 + T_{ax} + T_g) \int_0^l \left( \frac{\partial \phi_n(x)}{\partial x} \right)^2 dx \tag{3.43}$$

The gate induced tension $T_g$ results from the elongation of the CNT due to deflection and can be expressed as (see Appendix A.2):

$$T_g = \frac{EA_c}{2l} \int_0^l \left( \frac{\partial z(x,t)}{\partial x} \right)^2 dx = \frac{EA_c}{2l} z(t)^2 \int_0^l \left( \frac{\partial \phi_n(x)}{\partial x} \right)^2 dx \tag{3.44}$$

When inserting Equation 3.44 into Equation 3.43:

$$k_{tension} = \underbrace{(T_0 + T_{ax}) \int_0^l \left(\frac{\partial \phi_n(x)}{\partial x}\right)^2 dx}_{\text{Axial tuning}} + \underbrace{\frac{EA_c}{2l} z(t)^2 \left(\int_0^l \left(\frac{\partial \phi_n(x)}{\partial x} dx\right)^2\right)^2}_{\text{Gate bias tuning}} \tag{3.45}$$

The deflection is described by $z(t) = z_s + z_d \cos(\omega t)$, where $z_s$ represents the static component, which is influenced by the applied static gate bias $V_{GD}^{DC}$, and $z_d$ is the dynamic amplitude of oscillation. This static deflection is the mechanism through which the gate bias tuning method induces axial strain or tension in the structure. A detailed solution for $z(t)$ is presented in Section 3.4.

### 3.3.2 Effective Excitation Force

CNT resonators are typically actuated by applying a voltage between the CNT and an underlying gate electrode, generating an electrostatic force that drives mechanical motion. The interaction between the distributed electrostatic force and the mechanical mode of the resonator was captured through a modal projection, yielding the effective force acting on the $n^{th}$ vibrational mode as

$$F_{eff} = \int_0^l f(x,t)\phi_n(x)dx \tag{3.46}$$

where $l$ is the length of the CNT. This formulation highlights how the spatial profile of the driving force couples to the resonator's mechanical modes and provides a foundation for deriving a reduced-order, lumped-element description of the system dynamics.

Beginning with the wire-above-ground capacitance model, an expression for the distributed electrostatic force $f(x,t)$ is obtained, projected onto the relevant mode shape, and a final closed-form expression for $F_{eff}$ is derived that captures the dependence on geometry, bias voltage, and dynamic displacement.

The electrostatic actuation force acting on the resonator is modelled by considering the capacitance per unit length $\bar{C}_g(z)$ between the CNT and the underlying gate electrode, commonly modelled as [61]:

$$\bar{C}_g(z) = \frac{2\pi\varepsilon_0}{\cosh^{-1}\left(\frac{g_0 - z}{r}\right)} \approx \frac{2\pi\varepsilon_0}{\ln\left(2\frac{g_0 - z}{r}\right)} \quad \text{for} \quad g_0 \gg r \tag{3.47}$$

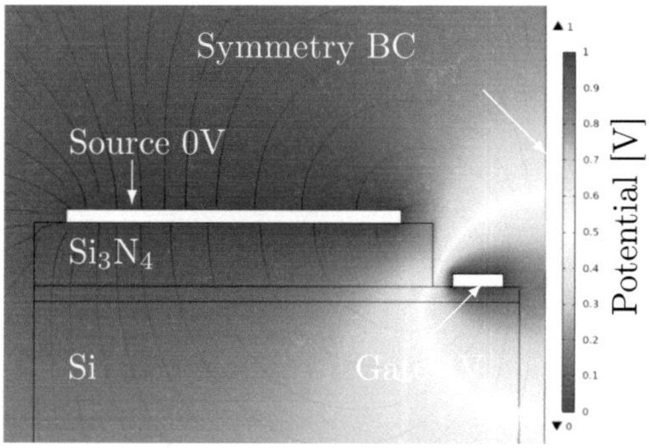


**(a)** Simulated electrostatic potential distribution obtained from a 2D COMSOL model, illustrating the shielding effect caused by the source and drain electrodes. A 1 V potential is applied to the gate electrode, while the source and drain are grounded. The effective potential near the CNT is visibly reduced toward the clamped ends, where shielding dominates.

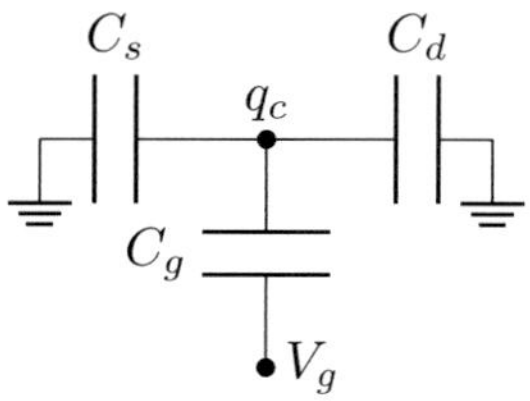


**(b)** Equivalent capacitance model representing the electrostatic coupling between the CNT and the surrounding electrodes. The gate capacitance $C_g$ is divided by the source and drain shielding capacitances $C_s$ and $C_d$, forming a capacitive voltage divider that reduces the net charge $q_c$ on the CNT.

**Figure 3.5:** Shielding effect of source and drain electrode.

where $g_0$ is the initial gap between the CNT and gate, and $z(t)$ is the dynamic displacement. This expression neglects fringing fields and is valid for long CNTs suspended above wide gate electrodes.

For short channel devices, the electrode shielding effect needs to be taken into account, which effectively reduces the electric field on the suspended CNT. Figure 3.5a shows how the effective potential applied by the gate electrode is significantly reduced towards the source/drain electrodes.

Shielding by the source/drain electrode can effectively reduce the applied force on the CNT. This effect becomes relevant for very short CNTs or large gate distances. The shielding effect can be modelled as shown in Figure 3.5b.

In the absence of electrode shielding, the charge $q_c$ induced on the CNT by the gate electrode is given by

$$q_c = C_g \cdot V_g \tag{3.48}$$

where $C_g$ is the gate-to-CNT capacitance, and $V_g$ is the applied gate voltage.

However, in practical devices, the presence of nearby source and drain electrodes introduces additional capacitances, which partially screen the electric field generated by the gate. Assuming the shielding capacitances are symmetric and denoted as $C_s = C_d$, they can be modelled as forming a parallel configuration with the gate, resulting in a capacitive voltage divider with $C_g$. The effective charge on the CNT is then reduced and can be expressed as:

$$q_c = \frac{2C_s \cdot C_g}{C_g + 2 \cdot C_s} V_g = \alpha_{elec} \cdot C_g \cdot V_g \tag{3.49}$$

where $\alpha_{elec} \in [0, 1]$ is a dimensionless correction factor quantifying the reduction in electrostatic actuation due to shielding.

The shielding capacitances effectively reduce the charge accumulated on the CNT, thereby lowering the electrostatic force exerted by the gate electrode. To estimate the magnitude of this shielding effect, a two-dimensional electrostatic simulation was carried out in COMSOL Multiphysics, following a similar approach as described by Muoth et al. [124]. In the simulation, a voltage of 1 V was applied to the gate electrode, while the source and drain electrodes were grounded. The resulting electrostatic potential along the CNT was extracted, and the spatial average in the suspended region was used to determine the force reduction coefficient $\alpha_{elec}$. This was done for all different gate distances $g_0$ and suspended lengths $l$ used in this work.

$$\alpha_{elec}(g_0 = 490\,\text{nm}, l = 2.9\,\mu\text{m}) = 0.4 \tag{3.50}$$

$$\alpha_{elec}(g_0 = 2\,\mu\text{m}, l = 2\,\mu\text{m}) = 0.47 \tag{3.51}$$

This value is specific to the geometry and gate-CNT separation and needs to be simulated for each different device architecture. Although $\alpha_{elec}$ generally also depends on the displacement of the CNT, this dependence is neglected in the current analysis due to the assumption of small vibrational amplitudes.

With this correction factor, the electrostatic actuation force applied to the CNT can be written as:

$$F = \frac{d}{dz}\left(\frac{\alpha_{elec}}{2}\bar{C}_g V^2\right) = \frac{\alpha_{elec}}{2}\frac{d\bar{C}_g}{dz}V^2 \tag{3.52}$$

The long simplification of the action force $F$ into $F_{eff}$ (from Equation 3.41) is skipped and the full deduction leading to this equation can be found in [66]:

$$F_{eff} = \frac{\alpha_{elec}}{2}V_{GD}^2(t)\bar{C}'\int_0^l \phi(x)^2dx + \underbrace{\frac{\alpha_{elec}}{2}V_{GD}^2(t)\bar{C}''\int_0^l \phi(x)^3dx \cdot z(t)}_{\text{Spring softening term}} \tag{3.53}$$

$$\bar{C} = \frac{2\pi\varepsilon_0}{\ln(\frac{2g_0}{r})^2} \qquad \bar{C}' = \frac{2\pi\varepsilon_0}{g_0\left(\ln(\frac{2g_0}{r})\right)^2} \qquad \bar{C}'' = \frac{2\pi\varepsilon_0}{g_0^2\left(\ln(\frac{2g_0}{r})\right)^3} \tag{3.54}$$

The total capacitance is then:

$$C_g = \bar{C} \cdot l \tag{3.55}$$

The electrostatic force for actuation will also cause a term in the equation which softens the stiffness of the CNT. The softening term is included in the model, however it is not relevant for CNT resonators (in contrast to graphene resonators [71]), as the capacitance between CNT and gate is small. For a typical CNT resonator, the gate capacitance is in the order of $1 \times 10^{-18}$ F (see Tab. 3.1), for graphene resonators, the gate capacitance is two orders of magnitudes larger [125]. Chen et al. [125] demonstrated the effect of spring softening.

The effect of spring softening is inherent of electrostatic actuation and has been used for the weak tuning of 3D bulk resonators. It is most prominent for 2D resonators.

### 3.3.3 The Explicit Duffing Equation for CNT Resonators

With the findings in the previous sections, Equation 3.42 can be rewritten in the form of the Duffing equation [83].

The effective stiffness $k_{eff}$ (Eqn. 3.40) was further investigated in Section 3.3.1. Its tension dependent term is split into three terms: potential pre-tension $T_0$, applied tension $T_{ax}$, and gate-induced tension $T_g$ (Eqn. 3.45). The base stiffness (first term of Eqn. 3.40) remained unchanged. In Section 3.3.2, the derivation of the effective force yielded a softening term (Eqn. 3.53). All of these terms are now used to rewrite $k_{eff} \cdot z(t)$ in Equation 3.42:

$$
\begin{aligned}
k_{eff} \cdot z(t) = z(t) \cdot \quad & EI_y \int_0^l \left( \frac{\partial^2 \phi_n(x)}{\partial x^2} \right)^2 dx && \text{beam stiffness} && (3.56) \\
+z(t) \cdot \quad & (T_0 + T_{ax}) \int_0^l \left( \frac{\partial \phi_n(x)}{\partial x} \right)^2 dx && \text{axial tension} && (3.57) \\
-z(t) \cdot \quad & \bar{C}_g'' V^2 \int_0^l \phi_n^3(x) dx && \text{spring softening} && (3.58) \\
+z(t)^3 \cdot \quad & \frac{EA_c}{2l} \left( \int_0^l \left( \frac{\partial \phi_n(x)}{\partial x} dx \right)^2 \right)^2 && \text{gate bias tension} && (3.59)
\end{aligned}
$$

The three terms originating from the beam stiffness, the axially applied tension, and the spring softening are all linear with $z(t)$ and are therefore summarized in the linear spring constant $k_0$:

$$
\begin{aligned}
k_0 = EI_y \int_0^l \left( \frac{\partial^2 \phi_n(x)}{\partial x^2} \right)^2 dx + (T_0 + T_{ax}) \int_0^l \left( \frac{\partial \phi_n(x)}{\partial x} \right)^2 dx \\
- \bar{C}_g'' V^2 \int_0^l \phi_n^3(x) dx \qquad (3.60)
\end{aligned}
$$

The stiffness term originating from the gate bias induced tension is multiplied with $z(t)^3$ and is referred to as the nonlinear spring constant $k_3$. Here it is important to remember, that $z(t)$ includes both the static displacement, that is mainly influenced by the gate bias tuning, and a dynamic displacement, mainly influenced by the gate AC component:

$$
k_3 = \frac{EA_c}{2l} \left( \int_0^l \left( \frac{\partial \phi_n(x)}{\partial x} dx \right)^2 \right)^2 \qquad (3.61)
$$

$F_{eff}$ was derived in Section 3.3.2 (Eqn. 3.53) and is also inserted in Equation 3.42, yielding the full Duffing equation:

$$m_{eff}\ddot{z}(t) + b_{eff}\dot{z}(t) + k_0 z(t) + k_3 z(t)^3 = \frac{\alpha_{elec}\bar{C}_g' V^2}{2} \int_0^l \phi_n^2(x)dx \quad (3.62)$$

The damping term $b_{eff}\dot{z}(t)$ represents the linear dissipation rate. For larger amplitudes of oscillation (nonlinear regime), the linear damping term can be extended with $\eta z(t)^2\dot{z}(t)$, where $\eta$ is the nonlinear damping coefficient, which takes material-dependent dissipation into account [83].

In this section, the general equation of motion was algebraically manipulated into the nonlinear ordinary differential equation also known as the Duffing equation. Explicit equations for all effective parameters were derived and can now be used further for approximating the solution of the Duffing equation with the method of harmonic balancing. Chapter 4 will explain how the axial tension $T_{ax}$ is applied with the straining platform.

## 3.4 Harmonic Balancing for the Duffing Equation

Harmonic balancing (HB) approximates periodic solutions with a truncated Fourier series, which consists of multiple harmonics (or orders). The higher the order of the ansatz, the more precise the solution. Whereas a $0^{\text{th}}$ order ansatz can only approximate a static displacement, the first-order ansatz can already capture eigenfrequencies accurately. A second-order ansatz allows for the approximation the first subharmonic and superharmonic component.

HB is particularly useful in the nonlinear regime, where it effectively models nonlinear spring hardening or softening effects with remarkable precision and low computational cost. The low computational cost of HB, compared to direct time-domain simulations, comes from several key factors. By transforming the system into a frequency-domain algebraic formulation, HB reduces the problem to a matrix equation. In the linear case, this system can be solved efficiently using matrix inversion techniques such as LU decomposition. For nonlinear cases, numerical solvers like Newton-Raphson are employed to iteratively solve the resulting algebraic system. The Newton-Raphson method for nonlinear algebraic systems is implemented in python as `fsolve` and is used in this work.

HB approximates periodic solutions such as:

$$z_h^{N_h}(t) = \sum_{n=0}^{N_h} z_{sn} \sin(n\omega t) + z_{cn} \cos(n\omega t) \tag{3.63}$$

where $N_h$ is the truncation order, $z_{sn}$ the respective coefficient of the $n^{th}$ sine harmonic, and $z_{cn}$ the respective coefficient of the $n^{th}$ cosine harmonic.

Ansatz $0^{\text{th}}$ order:

$$z_h^0(t) = z_{c0} \tag{3.64}$$

Ansatz $1^{\text{th}}$ order:

$$z_h^1(t) = z_{c0} + z_{s1} \sin(1\omega t) + z_{c1} \cos(1\omega t) \tag{3.65}$$

Ansatz $2^{\text{th}}$ order:

$$z_h^2(t) = z_{c0} + z_{s1} \sin(1\omega t) + z_{c1} \cos(1\omega t) + z_{s2} \sin(2\omega t) + z_{c2} \cos(2\omega t) \tag{3.66}$$

Alternatively, instead of using sine and cosine, it is possible to only use cosine and include a phase. The individual coefficients can be translated into more meaningful values like:

$$\text{Static discplacement:} \quad z_s = z_{c0} \tag{3.67}$$

$$\text{Dynamic amplitude:} \quad z_{d1} = \sqrt{z_{s1}^2 + z_{c1}^2} \tag{3.68}$$

The $0^{\text{th}}$ order term $z_{c0}$ is often neglected, which is correct if the actuation force does not contain a static term. For the case of CNT tunable resonators this term is crucial as this model needs to be capable of including the constant deflection by the gate bias tuning force. The two terms of each order can be combined into a single sine term with a dynamic amplitude $z_{dn}$ and a non-zero phase term.

The ansatz of the chosen truncation order is then inserted into Equation 3.62. By performing coefficient matching (equating the coefficients of $\sin(n\omega t)$ and $\cos(n\omega t)$ on both sides of the equation) the system of algebraic equations is obtained. Each equation corresponds to a specific harmonic and must be satisfied ("balanced") individually, hence the name harmonic balancing. The derivation of the system of equation for different truncation orders for the specific case of the CNT resonator are shown in Appendix A.3.

The HB therefore provides an approximation to the solution of the Duffing equation (Section 3.3.3). To model the dynamic behaviour of the CNT, a mechanical model of the CNT as continuum is required.

## 3.5 CNTs as Continuum Model

Carbon nanotubes (CNTs) are rolled up graphene sheets with their lattice structure being described in solid state physics using two basis vectors $\vec{a}_1$ and $\vec{a}_2$ with a length of of 0.246 nm and an inter-angle of 60°. The chiral indices $(n, m)$ are defined on the graphene lattice to specify the CNT perimeter.

The direction in which the CNT is rolled up is described in the chirality vector $\vec{C} = n\vec{a}_1 + m\vec{a}_2$ and fundamentally determines its electrical and mechanical properties. The largest influence of the chirality on the mechanical behaviour is through the radius of the CNT.

Two specific chiralities are so unique that they have distinct names: zigzag and armchair. The chiral vector of an armchair CNT is shown in Figure 3.6a, where $n = m$. Armchair CNTs exhibit metallic behaviour. Zigzag CNTs have a chiral vector, where $n = 0$ and can be either metallic or semiconducting. All other configurations are called chiral.

The mass of each carbon atom is $m_C = 12\,\mathrm{Da} = 1.9926 \times 10^{-26}\,\mathrm{kg}$ and the C-C bond length is $a_{CC} = 0.142\,\mathrm{nm}$. The unit cell of graphene contains two carbon atoms and has an area of $A_{\mathrm{cell}} = \frac{\sqrt{3}}{2} a_{CC}^2$ (one third of the area of a hexagon). The areal density of graphene $\rho_{\mathrm{graphene}}$ can be estimated as [93]:

$$\rho_{\mathrm{graphene}} = \frac{2 \cdot m_C}{A_{\mathrm{cell}}} = 7.6 \times 10^{-7}\,\mathrm{kg\,m^{-2}} \tag{3.69}$$

From the graphene lattice, its lattice constant $a$ is computed as $a = \sqrt{3} \cdot a_{CC} = 0.245\,95\,\mathrm{nm}$ and the radius can be computed as [126]:

$$r = \frac{a_1}{2\pi} \sqrt{n^2 + m^2 + nm} \tag{3.70}$$

Typical radii for SWCNTs lie between 0.7 nm and 1.5 nm. The radius of the rolled up graphene single-layer determines the linear mass density $\mu(r)$ of the CNT and depends on the radius $r$ of the CNT:

$$\mu(r) = 2\pi r \cdot \rho_{\mathrm{graphene}} \tag{3.71}$$

This linear mass density describes a pristine SWCNT without any residues or adsorbates on the CNT surface. In reality, if the CNT is not permanently stored in perfect vacuum conditions, particles and other molecules will adsorb on the surface, increasing the linear mass density, without changing the stiffness of the CNT. The increased mass of the CNT will effectively reduce the eigenfrequencies and is referred to as

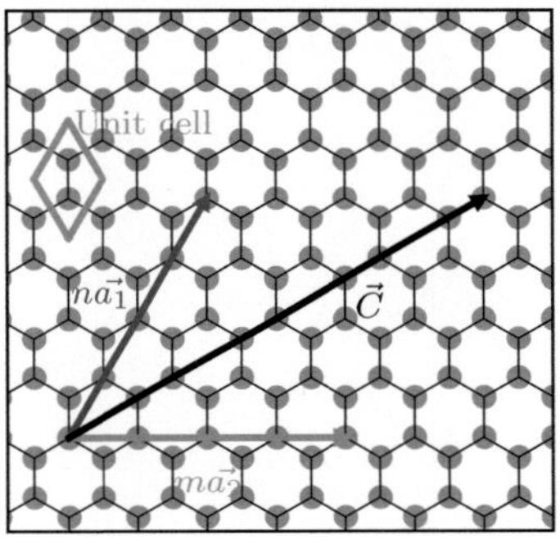


(a) The 2D hexagonal carbon lattice (graphene) is shown with its unit cell (red) and the lattice vectors $\vec{a}_1$ and $\vec{a}_2$. The chiral vector $\vec{C} = n\vec{a}_1 + m\vec{a}_2$ defines how the graphene sheet is rolled to form a CNT.

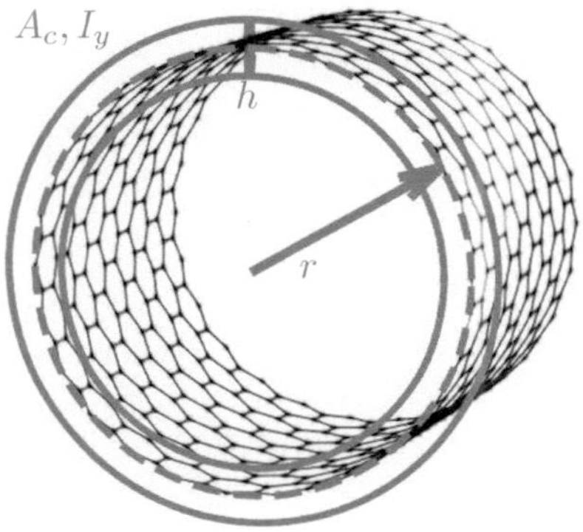


(b) Cross-sectional model of a carbon nanotube (CNT) in the continuum approximation. The CNT is modeled as a thin-walled cylindrical shell with radius $r$, wall thickness $h$, cross-sectional area $A_c$, and second moment of area $I_y$. This abstraction enables mechanical analysis using beam theory.

**Figure 3.6:** Continuum model of the CNT. The transition from the atomistic graphene lattice to a cylindrical shell model allows the use of classical continuum mechanics to describe the mechanical behaviour of carbon nanotubes.

mass-loading. The effective increase of linear density was not measured and is introduced as a fit-parameter called mass-loading factor $\alpha_{ML}$.

The usage of such mass-loading factor is common in MEMS/NEMS resonator systems and was applied for $MoS_2$ [127], graphene [128] and CNT resonators [129] and can exceed a factor of ten. To include the effect of mass loading, the linear density of a CNT is described as:

$$\mu(r) = 2\pi r \cdot \alpha_{ML} \cdot \rho_{\text{graphene}} \tag{3.72}$$

The continuum model is shown in Figure 3.6b where the radius $r$ and the wall thickness $h$ are defined. This continuum model has a cross section:

$$A_c = \pi \left(r + h/2\right)^2 - \pi \left(r - h/2\right)^2 = 2\pi \cdot r \cdot h \tag{3.73}$$

and a moment of inertia $I_y$:

$$I_y = \frac{\pi}{4} \left( (r + h/2)^4 - (r - h/2)^4 \right) \approx \pi r^3 h \tag{3.74}$$

The radius $r$, mass density $\mu(r)$, cross-sectional area $A_c(r, h)$, and moment of inertia $I_y(r, h)$ presented are all part of the tube model that is employed in this work for CNTs in contrast to the solid cylinder (rod) model. They are used to compute the effective parameters for the Duffing equation and reduce the unknown parameters down to the radius, the pre-tension and the mass-loading factor.

## 3.6 Modelling Results for a Typical CNT

In this section, the lumped Duffing equation is solved with a HB approach of the second order for typical CNT geometries and properties to investigate the resonator response to controllable external influences like the gate bias $V_{GD}^{DC}$, the total strain of the CNT $\epsilon_i$, and the gate AC driving voltage $V_{GD}^{AC}$.

Figure 3.7a shows the capability of the HB solution to capture the influence of the gate bias tuning effect. The eigenfrequency of the modelled CNT increases with larger gate biases as expected and demonstrated before.

Figure 3.7b shows the influence of applied strain $\epsilon_i$ (equivalent to the applied tension $T_i = EA\epsilon_i$) on the CNT. Here, the eigenfrequency rises, but the amplitude of oscillation decreases with higher strain.

The last external parameter that can be influenced is the driving voltage $V_{GD}^{AC}$. Its influence is shown in Figure 3.7c. The driving voltage directly increases the dynamic amplitude until the nonlinear regime is triggered. For large driving amplitudes, the jumps between the stable solutions become visible.

To compare the impact of gate bias tuning and axial strain tuning, both parameters are swept at the same time and the eigenfrequency is investigated in Figure 3.8.

The gate bias tuning has a large influence for lower axial tensions. With increasing the axial tension, the gate tuning effect decreases. Typical gate voltages only go up to 10 V or even less, confirming the limitation of this tuning method. The axial strain however shows a much larger influence on the eigenfrequency and can go much higher, reaching values of 335 MHz with tensions of 8.1 nN ($\epsilon = 0.4\,\%$).

With the tube model and the HB solution of the Duffing equation, the location of the eigenfrequencies will be modelled and the input parameter radius $r$, mass-loading factor $\alpha_{ML}$, and pre-strain $\epsilon_0$ will be used for fitting. The amount of fitting parameters is reduced to an absolute minimum for our case, because no RBM information of the radii was available. This improves the modelling presented in literature, where the cross-sectional area and the moment of inertia have often been fitted individually. The solving procedure is summarized in Figure 3.9.

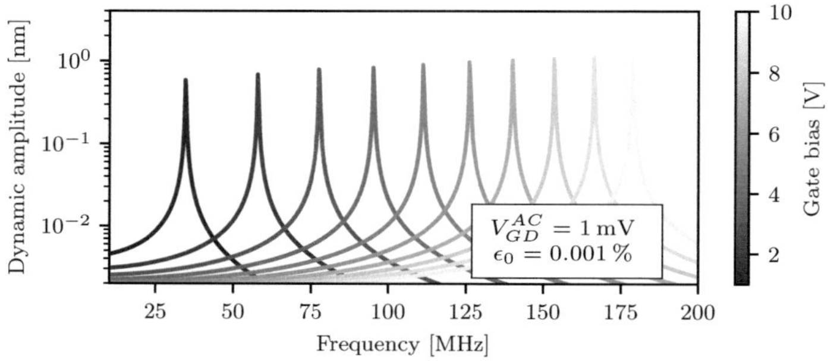


**(a)** Resonance tuning via gate bias. Dynamic amplitude spectra for increasing electrostatic gate voltages, showing a significant upshift in resonance frequency with higher bias levels.

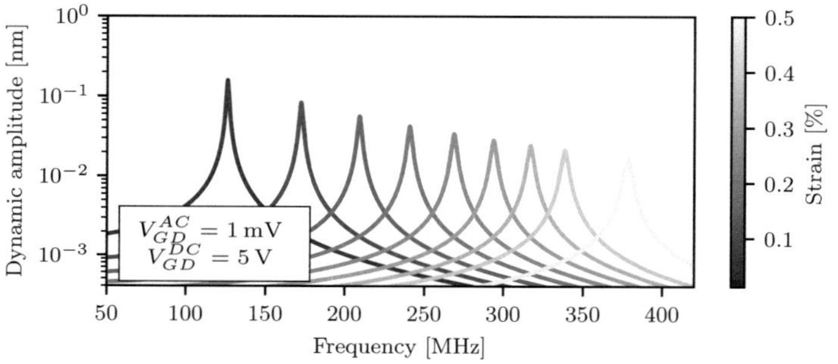


**(b)** Resonance tuning via axial strain. Frequency response curves for increasing axial strain, showing a strong stiffening effect and corresponding frequency upshift. The dynamic amplitude decreases with higher strains.

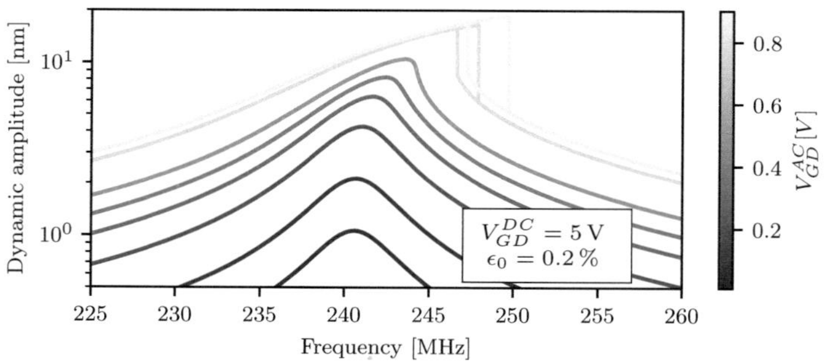


**(c)** Nonlinear dynamic response under increasing AC actuation voltage. The frequency response shows the onset of a Duffing-type nonlinearity for higher actuation voltages. The response transitions from symmetric Lorentzian peaks to asymmetric, bent profiles characteristic of stiffening nonlinearity.

**Figure 3.7:** Simulated dynamic amplitude spectra under different tuning mechanisms. (a) and (b) compare resonance tuning effects from gate-induced electrostatic tuning and axial strain-induced tuning. (c) illustrates the emergence of nonlinear Duffing behaviour for increasing AC actuation voltage.

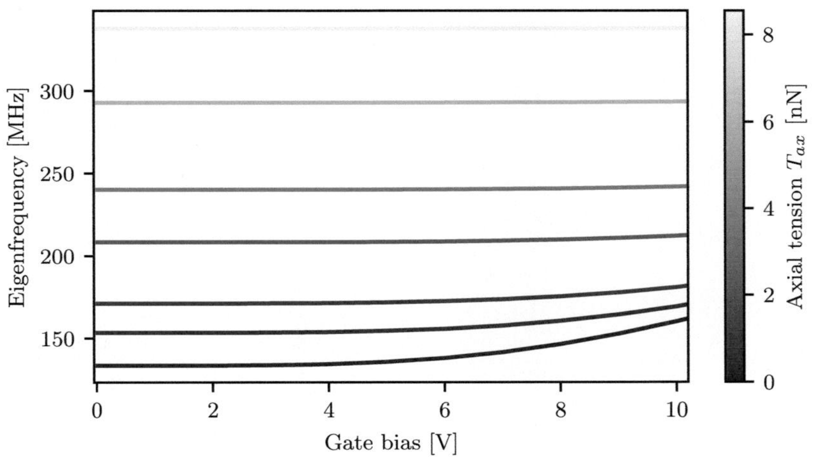


**Figure 3.8:** Comparison of gate-induced and strain-induced tuning of the fundamental mode. The plot shows the eigenfrequency of the first flexural mode as a function of gate bias, for different levels of applied axial tension $T_{ax}$. Both gate bias and axial strain contribute to resonance tuning via tension-induced stiffening. However, the effect of axial strain is significantly stronger, enabling a much broader tuning range, while the gate bias induced tension only shows a visible effect for lower axial tension levels.

**Table 3.1:** Typical CNT values used throughout this chapter for figures and visualization. Mode shape parameters and effective parameters are derived for a tension $T = 60\,\text{pN}$.

| | | Value | Reference |
|---|---|---|---|
| **Geometry** | Length $l$ | $2\,\mu\text{m}$ | Example |
| | Gate distance $g_0$ | $500\,\text{nm}$ | Example |
| | Radius $r$ | $1\,\text{nm}$ | Example |
| | Wall thickness $h$ | $340\,\text{pm}$ | Example |
| | Quality factor $Q$ | 100 | Example |
| **External** | Young's modulus $E$ | $1\,\text{TPa}$ | Common value |
| | Graphene density | $7.6 \times 10^{-7}\,\text{kg}\,\text{m}^{-1}$ | Eqn. 3.69 |
| | Tension $T$ | $60\,\text{pN}$ | Example |
| | Strain $\epsilon$ | $0.003\,\%$ | $T/E/A$ |
| | $C_0$ | $6.4 \times 10^{-18}\,\text{F}$ | Eqn. 3.55 |
| | $C_g'$ | $-9.3 \times 10^{-7}\,\text{F}\,\text{m}^{-2}$ | Eqn. 3.54 |
| **Tube model** | Mass-loading factor $\alpha_{ML}$ | 1 | pristine CNT |
| | Linear density $\mu$ | $4.77\,\text{kg}\,\text{m}^{-1}$ | Eqn. 3.71 |
| | Mass $m$ | $9.6 \times 10^{-21}\,\text{kg}$ | $\mu \cdot l$ |
| | Cross-section $A_c$ | $2.1 \times 10^{-18}\,\text{m}^2$ | Eqn. 3.73 |
| | Moment of inertia $I_y$ | $1.1 \times 10^{-36}\,\text{m}^4$ | Eqn. 3.74 |
| **Mode shape** | $\int_0^l \phi_n(x)dx/l$ | 0.56 | converges |
| | $\int_0^l \phi_n^2(x)dx/l$ | 0.43 | converges |
| | $\int_0^l \phi_n^3(x)dx/l$ | 0.36 | converges |
| | $\int_0^l \left(\frac{\partial \phi_n(x)}{\partial x}\right)^2 dx \cdot l$ | 4.87 | converges |
| | $\int_0^l \left(\frac{\partial^2 \phi_n(x)}{\partial x^2}\right)^2 dx \cdot l^3$ | 231.3 | diverges |
| **Duffing** | $m_{eff}$ | $4.08 \times 10^{-21}\,\text{kg}$ | Eqn. 3.38 |
| | $b_{eff}$ | $8.96 \times 10^{-15}\,\text{kg}\,\text{s}^{-1}$ | Eqn. 3.39 |
| | $k_0$ | $3.1 \times 10^{-5}\,\text{N}\,\text{m}^{-1}$ | Eqn. 3.60 |
| | $k_3$ | $3.2 \times 10^{12}\,\text{N}\,\text{m}^{-3}$ | Eqn. 3.61 |

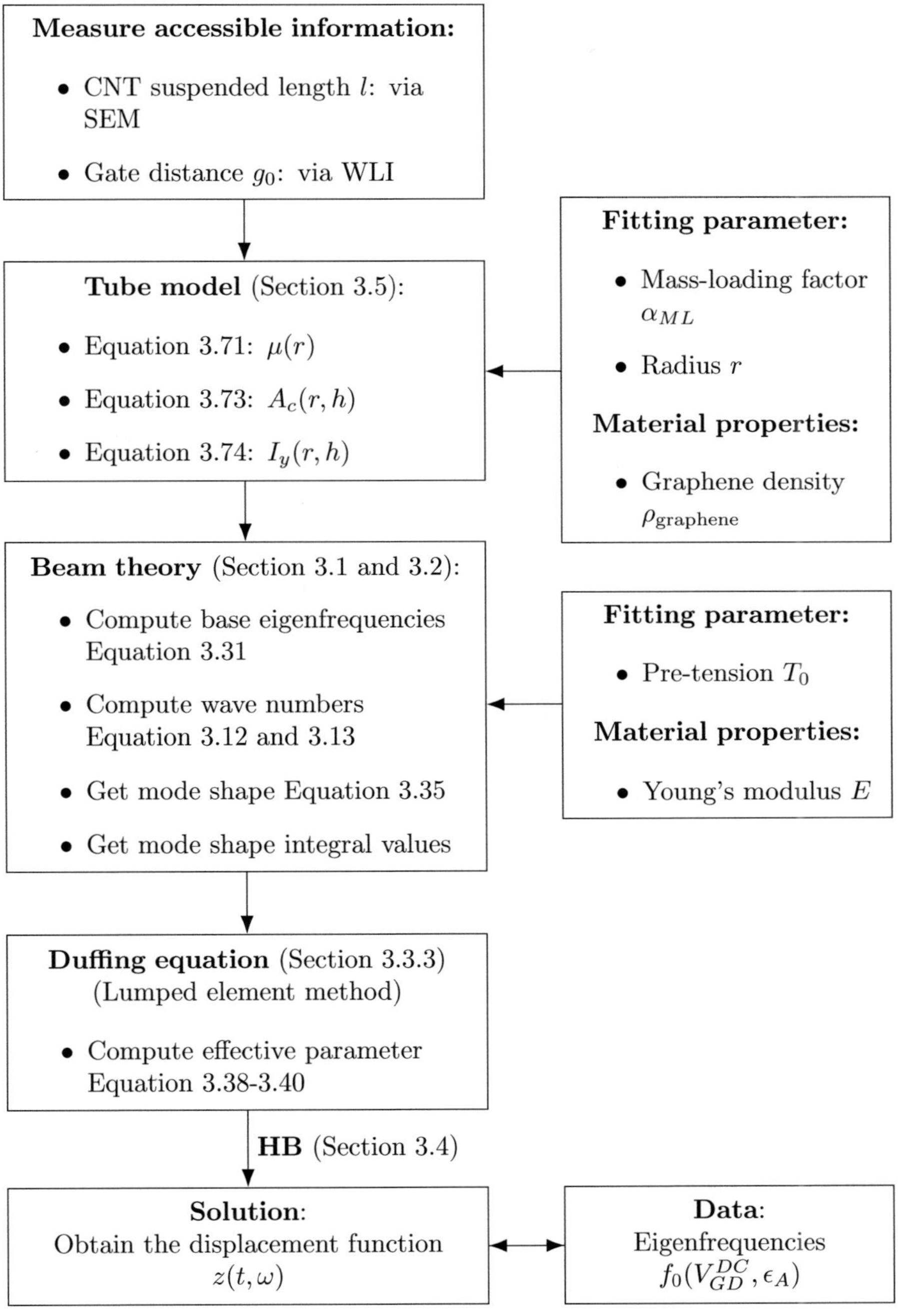


**Figure 3.9:** Flow chart of the solution approach from this Chapter.

# 4 The Mechanical Straining Platform

The importance of a MEMS straining platform was elaborated on in the introduction (Section 2.2). The limitation of gate bias tuning alone prevents the exploration of the full range of tunability of CNT resonators and the ability of slack removal requires the design of new approaches.

In this chapter, various design considerations for a specific type of axial strain tuning platform, the thermal actuator, are presented. Special attention is given to the geometry of the chevron actuator structure, the design of the separating trench between the movable electrodes, and the mitigation of internal stress that arises between different material layers during fabrication.

Following the design considerations, the fabrication process is described in detail. Several challenges associated with cleanroom processing are addressed. These include the etching of a narrow and deep trench into the silicon substrate, as well as the issue of delaminating electrodes due to accumulated stress or poor adhesion. Resolving these problems required careful process tuning and extensive iterative development throughout the fabrication phase.

After the successful fabrication of the MEMS platform, the characterization of the actuators is carried out. A key aspect is the current-displacement relationship, which is measured directly inside a scanning electron microscope. This approach enables highly precise quantification of displacement and allows the evaluation of actuator performance under controlled electrical input conditions.

To gain a deeper understanding of the actuator behaviour, particularly under high-temperature operation, simulations are performed using COMSOL Multiphysics. These simulations help analyse the thermomechanical response of the actuator structures and support the interpretation of experimental results. They also provide insight into the material behaviour beyond the intended operational range.

The design features and the working principles are discussed in Section 4.1, followed by the fabrication methods 4.2, the actuator characterization 4.3,

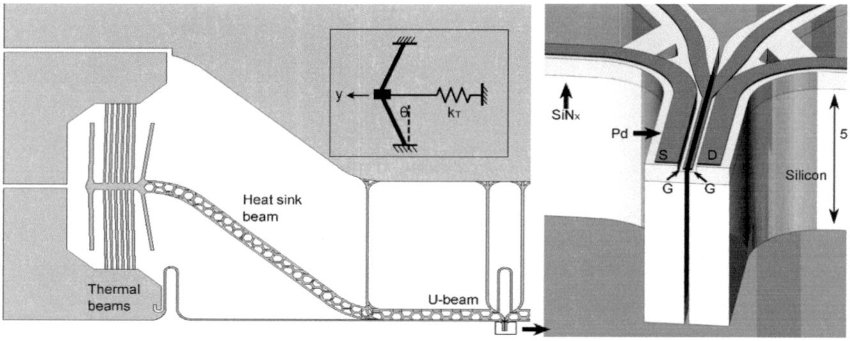


**Figure 4.1:** Previously presented design from the Micro- and Nanosystems group [74]. The thermal actuator is connected through a beam to separated electrodes, allowing for their relative displacement. Open Access figure from [74] published under open access under CC BY licence (Creative Commons Attribution 4.0 International license).

a FEM simulation to model and understand the behaviour 4.4. Then the results are summarized and discussed in Section 4.5.

## 4.1 Design of the MEMS Straining Platform

The design of the MEMS straining platform presented in this work builds upon earlier developments within the Micro- and Nanosystems group. The first thermal actuators were conceived and fabricated by M. Muoth, S.-W. Lee, and S. Truax et al. [72,73] and was then further developed by L. Kumar, L. V. Jenni et al. [66, 74]. Figure 4.1 shows a schematic of the previously presented MEMS platform as example for axial tuning for CNT resonators in Figure 2.6c, which already incorporated several key design considerations. These design aspects served as the foundation for the platform developed in this thesis and were further adapted and advanced to meet the specific requirements of CNT straining.

The design consists of a split gate electrode which allows for relative displacement between source and drain electrodes. Each electrode is connected by a long beam to a chevron structure. The chevron structure heats up during operation, which could affect the CNT as it is sensitive to temperature. Therefore, the heat dissipation path between actuator and electrodes need to be considered well.

The temperature influence includes thermal expansion/contraction of the CNT, changes in the electrical conductivity of the CNT itself as well as an influence on the contacts to the electrodes. Further, the mechanical properties such as the CNT stiffness, wall thickness or radius can be influenced by temperature [130, 131]. As these influences are complex and potentially different from tube to tube, the temperature at the CNT should remain constant to avoid influence and isolate measured effects purely to the applied strain. In the previous designs, the heat sink beam is rather long, which increases the heat resistance between the thermal actuator and the CNT. The beam is patterned with a hexagonal structure for mechanical stability and supported with springs, which anchor the beam to the rest of the substrate and also work as heat dissipation paths.

This thermal actuator design was studied in [74] and an optimal design was presented, which took both deflection force and deflection efficiency into account. In particular, the incline of the actuator beams was investigated (see sketch in Figure 4.1, inset). The optimized designs of the thermal actuators remain unchanged in this work and their working principle are further discussed in Section 4.1.1.

Both sides of the electrodes are connected with an U-shaped spring which restricts the deflection of both sides to the same plane. Due to the free-standing structures of the MEMS, internal layer tensile stress can cause the structure to bend upwards. This internal layer stress origins from silicon nitride, which was deposited at high temperatures. The out-of-plane bending of the electrodes posed a problem in the previous actuator designs. To counteract this effect, three counter measures were implemented on the design level and were added in the actuator design presented in this thesis.

1. The previous connection between the actuator and the electrodes was anchored only on one side. For this structure, if there was stress in the one-sided anchor, the structure would not just bend up, but in addition also tilt, which poses a problem for the CNT transfer (see Section 4.2.9). In this thesis, the tilting is reduced by using symmetrical springs on both sides of the connecting beam. In order to achieve this together with necessary space for the dry transfer, the actuators as well as the connecting beams were inclined (see Fig. 4.2).

2. The second counter measure is the removal of excess silicon nitride, wherever it is not needed to isolate the electrodes from the silicon device layer. The design in this work removes the silicon nitride on the "heat sink beams" between the electrodes and the thermal actuator. The deposition of silicon nitride is explained in Section 4.2.1.

3. The U-spring which forces the two electrodes into the same plane remain unchanged, as this was a measure already implemented in the previous designs to counteract out-of-plane bending (Fig. 4.1).

**Table 4.1:** Design parameter variations of the thermal actuator.

| Design variation | | | |
|---|---|---|---|
| # Beams | 6 | 9 | 13 |
| # Springs | 0 | 1 | 2 |
| Inclination angle | 30 | 35 | 40 |

Figure 4.2 shows one MEMS straining platform from this work. The necessary space for the CNT transfer is highlighted as well as the inclination of the beams and thermal actuators. Wire-bonds are shown.

Other design features are heat capacitances shielding each side of the chevron structure to increase the temperature in the the thermal actuator. The number of actuator in the chevron structure is varied, as well as the number of restoring sprigs and and angle of inclination for the beam.

### 4.1.1 The Thermal Actuators

A schematic sketch of the working principle is included in Figure 4.1 [74]. This study includes the analysis of expected displacement due to thermal expansion coefficient $\alpha$, temperature increase $\Delta$T and geometry of the beam. In particular, this study investigated an optimal angle of the beam of the thermal actuators.

Thermal expansion in the actuator relies on the Joule heating effect, whereby electrical energy is converted into heat due to resistive losses. The power dissipated in the actuator is given by

$$P = I_{act}V = I_{act}^2 R_{act} \tag{4.1}$$

where $P$ is the dissipated power, $I_{act}$ is the current through the actuator, $V$ the applied voltage, and $R_{act}$ the electrical resistance of the actuator. Due to the temperature increase of the dissipated energy, the silicon beams of the chevron structure expand, move the shuttle, and displace the electrodes relative to each other by $\Delta l$.

Source and drain are connected by an inclined beam to the chevron structure. In this study, three different designs of the actuator are considered using six, nine and thirteen beams (Figure 4.3). The thermal actuators are driven simultaneously with the same driving current which defines their displacement. To exclude the influence of varying contact resistances due to positioning or quality of wire-bonds, driving currents were chosen instead of voltages.

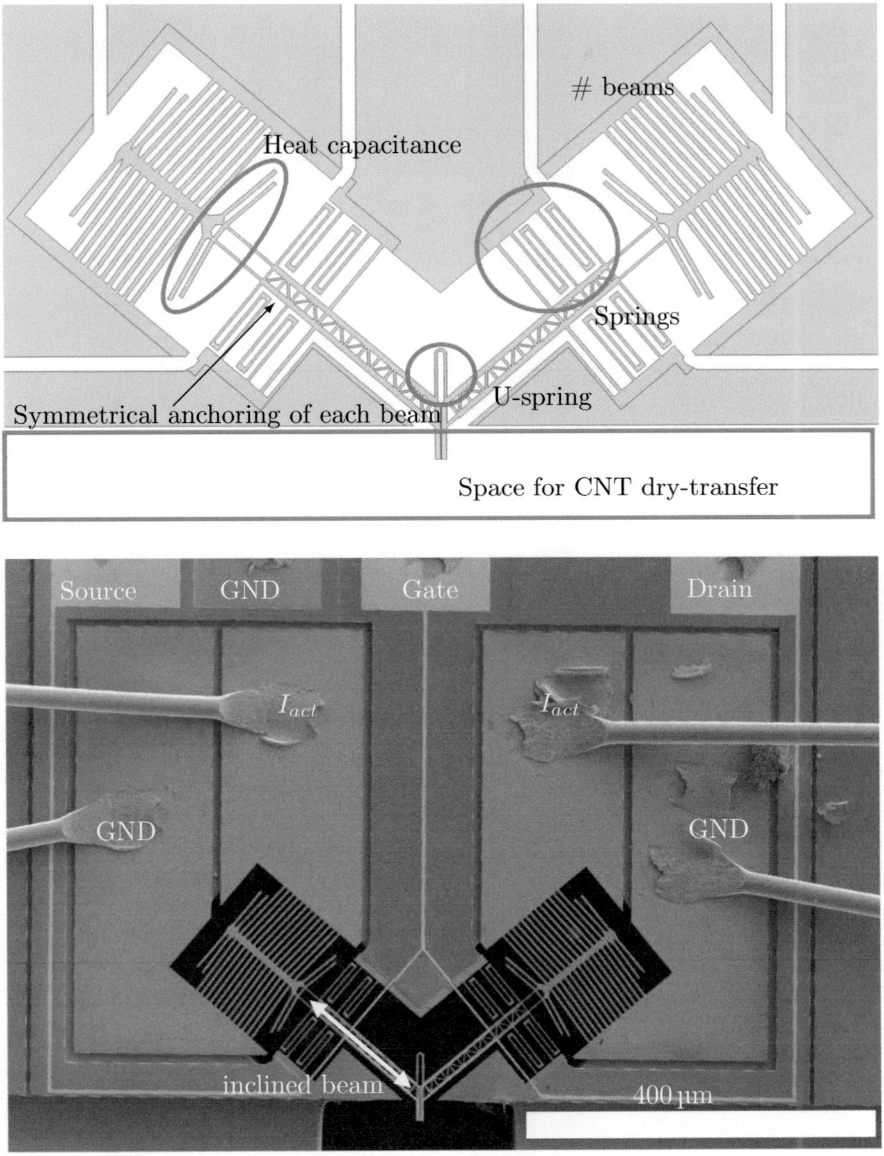


**Figure 4.2:** SEM image and schematic of the full MEMS straining platform. The schematic (top) highlights key design elements including the thermal actuators, U-spring, inclined beam, symmetrical anchoring, and the central region designated for CNT dry-transfer. The SEM image (bottom) shows the fabricated device, with labelled electrodes (Source, Gate, Drain, GND) and actuation currents $I_{act}$ applied to the thermal actuator arms.

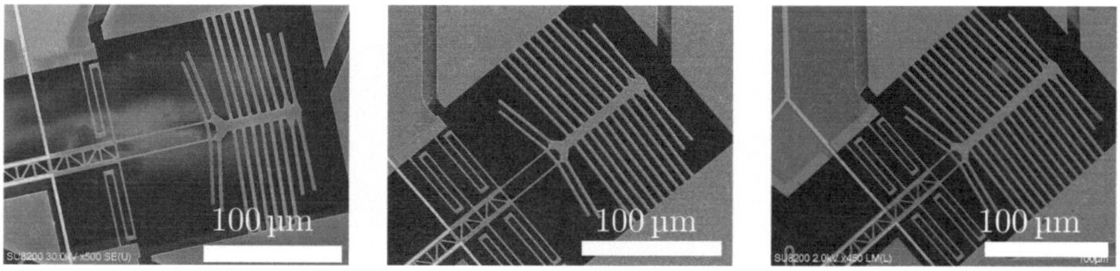


**(a)** A6 design. **(b)** A9 design. **(c)** A13 design.

**Figure 4.3:** Different thermal actuator designs with 6 (A6), 9 (A9), and 13 (A13) beams.

To achieve efficient operation and design, as much heat as possible should remain in the chevron structure, ensuring high temperatures and therefore higher displacement at same currents. The chevron structure is shielded with two heat capacitances. The actuator is connected to the electrodes with two 2 µm thin beams. These thin structures serve as heat resistances to reduce the heat flowing out of the thermal actuator. The rest of the beam functions as an undesired heat capacitance and is anchored with the folded springs which serve as additional heat dissipation paths.

Heating up the electrode regions where the CNT is, should also be avoided. This is due to the several temperature dependent properties, such as thermal expansion which could influence sensitive strain measurements and electrical conductance which might change current amplitudes. The description of the temperature dependent quantities is presented in further detail in Section 4.4.1.

### 4.1.2 Electrode Design

The electrodes of the straining platform need to be separated in the middle in order to be displaced relative to each other. An SEM picture is shown in Figure 4.4a. Both sides of the electrodes are connected to the inclined beam and by an U-shape spring. Different angles of the inclined beam were designed (30°, 35°, 40°). However, this design feature showed no effect.

Two electrode designs were fabricated and successfully measured on devices (with CNTs, showing measurable resonance behaviour). The designs were as follows: one design contained a wider source/drain electrode pair and shorter suspended CNT length, and the other had a longer suspended length and narrower source/drain electrodes (geometry shown in Tab. 4.2, schematic shown in Fig. 4.4b). The different suspended lengths of CNTs showed the expected influence on the eigenfrequencies (see Section 3.6). The width of the electrodes was varied to investigate influences of the contact region

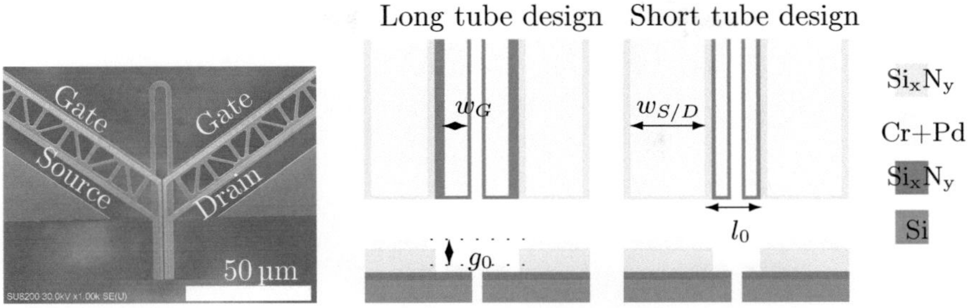


**(a)** SEM image of the electrodes.

**(b)** Designs for the long and short suspended CNT. The $Si_xN_y$ has been deposited in one step. The different colouring highlights the gate distance and is the same colouring as used in the fabrication Section 4.2.

**Figure 4.4:** Electrodes for the tunable CNT resonator.

**Table 4.2:** FET geometries.

| Value | short device | wide device |
|---|---|---|
| Suspended length $l_0$ | 2.9 nm | 4.2 nm |
| Source/Drain width $w_{S/D}$ | 2.6 µm | 2 µm |
| Gate width $w_G$ | 400 nm | 800 nm |
| Trench width | 400 nm | 400 nm |

between the CNT and the electrode on the contact strength and the contact resistance. However, no influence could be confirmed, in part also due to the small number of working devices.

The electrodes were fabricated using palladium, because palladium's work function ($\approx 5.1$ eV) matches the CNTs work function closely. It also yields lower contact resistances than other materials for example gold.

The design of the anchoring springs was adapted in comparison to the previously researched design [66, 74]. Instead of having one-sided straight beams, the anchoring is now two-sided for higher symmetry and the beams are folded, which allows for the release of internal layer stress. Design variations included zero, one or two springs on each side, however the amount of springs did not show any influence on the current-displacement curves.

## 4.2 Fabrication of the MEMS Straining Platform

The fabrication process is based on previous work by Dr. Laura Vera Jenni [132,133]. The fabrication of the CNT resonator can be divided into different steps:

1. Fabrication of the receiving MEMS substrate
2. Wire-bonding and characterization of the MEMS substrate
3. CNT dry transfer process and post-treatment

The fabrication of the MEMS substrate includes the following sequence: low-pressure chemical vapor deposition (LPCVD) of silicon nitride (L0), marker definition, silicon nitride patterning (L1), gate-level patterning (L2), electrode deposition (L3), device layer patterning for the thermal actuator structure (L4), mesa etching for dry transfer compatibility (L5), and finally backside etching and structure release (L6) (all summarized in Fig. 4.6). The process combines UV lithography and e-beam as high accuracy alignment is necessary below the usual achievable manual alignment precision of around 500 nm when using vernier structures.

After completing the wafer-level processing (Fig. 4.5), individual devices are separated from the wafer, packaged, and wire-bonded and then tested for shorts and functionality. Devices are then tested for electrical continuity and functionality. The final fabrication step consists of the CNT dry transfer, accompanied by established pre-treatment and post-treatment procedures to ensure CNT adhesion and electrical contact quality.

The initial LPCVD silicon nitride deposition (L0) was performed by the Center of MicroNanoTechnology (CMI) at EPFL Lausanne. All subsequent fabrication steps were carried out at the Binnig and Rohrer Nanotechnology Center (BRNC) at IBM Research in Rüschlikon.

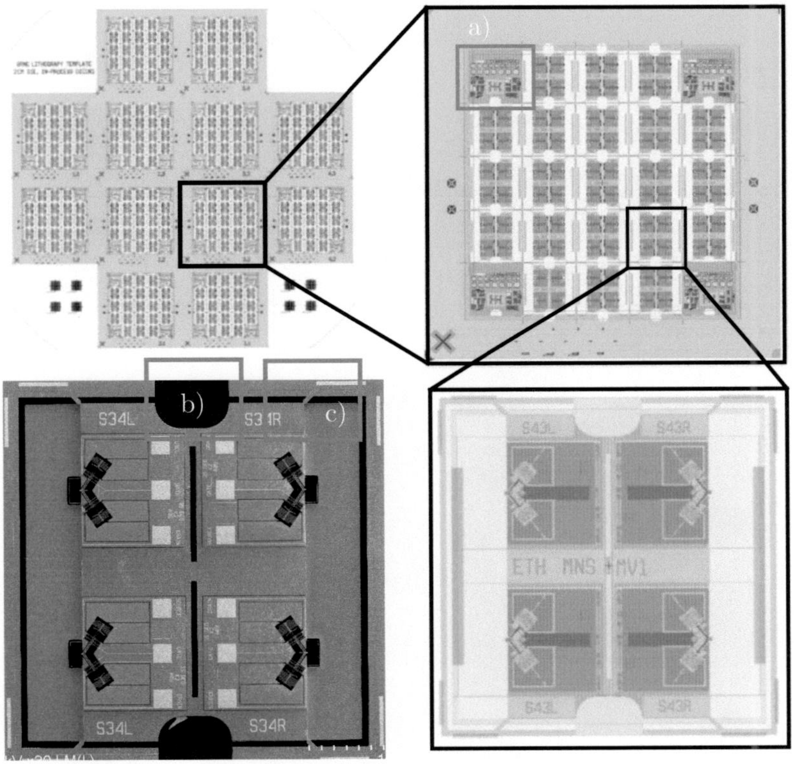


**Figure 4.5:** The BRNC wafer level template for e-beam fabrication consists of 12 individual chips, each with individual e-beam alignment markers for potential dicing. Each chip consists of 25 cells. 4 cells in the corners have test structures on them, 21 chips carry different substrate designs. These cells have 2 sides of devices that can be broken out from the wafer. Tweezers holes (b) allow to break the small bridges (c) to separate the devices for packaging and wire-bonding.

**(a)** L1: $Si_xN_y$ removal via RIE to expose Si for wirebonding of actuators and reducing bending.

**(b)** L2: $Si_xN_y$ gate patterning via RIE.

**(c)** L3: Palladium electrode deposition.

**(d)** L4: RIE and DRIE beam shape patterning.

**(e)** L5: Mesa etch with BHF and DRIE.

**(f)** L6: Backside removal with DRIE and BHF release.

**Figure 4.6:** Fabrication process overview.

### 4.2.1 L0: Wafer Preparation and Marker Deposition

The fabrication process uses a commercially available silicon-on-insulator (SOI) wafer, consisting of a highly doped 5 µm silicon device layer (resistivity $< 0.02\,\Omega\,\text{cm}$) a 2 µm $SiO_2$ insulating layer, and a 400 µm silicon handle layer. The thermal actuators are fabricated from the device layer, which necessitates electrical isolation of the CNT FET electrodes. Silicon nitride ($Si_xN_y$ ) was selected as the insulating material, following historically established fabrication protocols [133].

LPCVD silicon nitride is deposited across the entire wafer surface, including regions that will later form free-standing structures. However, since internal layer stress can lead to structural deflection, the silicon nitride is removed from those regions, as well as from areas designated for wire bonding.

The internal stress of silicon nitride is highly sensitive to the deposition method. In this work, both low-pressure chemical vapour deposition (LPCVD) and plasma-enhanced chemical vapour deposition (PECVD) were explored. Low-stress silicon nitride can be achieved using in-house PECVD processes at the Binnig and Rohrer Nanotechnology Center (BRNC) [133]. However, PECVD-deposited silicon nitride was observed to have significantly higher etch rates in buffered hydrofluoric acid (BHF), which is used during the critical release step (see Section 4.2.7), resulting in delamination of the electrodes. This risk was mitigated by switching to LPCVD-deposited silicon nitride.

The LPCVD process was carried out at the Centre of MicroNanoTechnology (CMI) at EPFL Lausanne. It included a critical cleaning procedure known as RCA (developed by the Radio Corporation of America [134]), which consists of organic and particle removal, a potential oxide strip using HF, and an ionic clean. This pre-treatment is essential prior to high-temperature deposition to ensure both uniform film growth and strong adhesion. The deposition of 500 nm $Si_xN_y$ was performed using a DCS/$NH_3$ gas ratio of 4:1 at 825 °C. Unlike the standard stoichiometric silicon nitride ($Si_3N_4$, with an internal stress $\sigma_i = 1150\,\text{MPa}$), the deposited film was intentionally low-stress and non-stoichiometric. The expected internal stress for LPCVD-deposited $Si_xN_y$ is approximately 250 MPa.

Following the LPCVD step, all subsequent fabrication processes were conducted at the BRNC. Marker structures (including both the BRNC template and manually designed alignment markers) were deposited using a double-layer PMMA/MMA lift-off process, in accordance with Dr. Jenni's process using AR-P 617.06 and AR-P 672.02 resists. Exposure was ordered at BRNC. After exposure, development was carried out using a 2:1 mixture of methyl isobutyl ketone (MIBK) and isopropanol (IPA) for 2 min, followed by a 30 s rinse in IPA.

To facilitate optical wafer alignment, three pairs of alignment markers were placed on each side of the wafer. This layout ensures that subsequent alignment steps do not degrade the visibility or accuracy of the markers across six required optical exposures. The markers were adapted from a previous design and include Vernier structures for enhanced alignment precision (see Appendix C.7). Locating the markers at the wafer edges improves alignment accuracy compared to alignment on individual chips.

Metal marker deposition was performed using an EVATEC electron-beam evaporator. A 0.5 nm chromium layer was used for adhesion, followed by a 60 nm gold layer. Lift-off was carried out in acetone using an ultrasonic bath (USB).

### 4.2.2 L1: Silicon Nitride Removal

The next step is the removal of the $Si_xN_y$ layer. This is performed to reduce the impact of thin-film stress and is additionally required to clear the regions around the contact pads of the thermal actuators, which are later used for direct wire bonding to the silicon device layer (see Section 4.2.8).

To define the etching areas, AZ 1512 photoresist is patterned using optical alignment and UV lithography. The mask layout is designed to tolerate typical misalignments on the order of 1 µm. For applications requiring higher accuracy, the bending of delicate device structures can be further minimized by introducing an additional electron-beam lithography step. This allows for more precise patterning and selective removal of silicon nitride but comes at the cost of increased fabrication time and complexity.

The $Si_xN_y$ layer is etched using reactive ion etching (RIE) for 23 min, ensuring sufficient over-etching to fully clear the target regions. After the etching step, the photoresist is removed using an acetone bath followed by an oxygen plasma clean (600 W, 3 min).

The complete fabrication process, including exact timings and parameters, is documented in Appendix C.1.

### 4.2.3 L2: Gate Patterning

The role of the gate distance $g_0$ in determining both the effective electrostatic actuation force and the tunability of the CNT resonator is discussed in detail in Section 3.3.2. The gate patterning step defines this critical distance by etching into the $Si_xN_y$ layer. High alignment precision is essential at this stage, as the gate trench must be centered accurately between the electrodes that will be deposited in subsequent steps.

Electron-beam lithography provides the required alignment accuracy, with typical positioning precision around 20 nm.

To achieve both high-resolution patterning and process efficiency, the "mix-and-match" photoresist AR-N 7520.17 is used. This resist behaves as a positive tone under electron-beam exposure and as a negative tone under UV exposure. This dual sensitivity allows non-critical areas to be patterned using UV lithography, while critical regions such as the gate trench are exposed with e-beam lithography, thereby optimizing alignment accuracy where needed.

After exposure, the resist is developed in AZ 300-47 for 4 min. The gate trench is then etched into the silicon nitride using reactive ion etching (RIE), with the etch time determining the final gate distance $g_0$. Since the etch is time-controlled, this parameter can only be varied on a wafer-to-wafer basis. For the actuators presented in this work, typical etch times ranged from 13 to 16 min, resulting in gate distances between 400 nm and 500 nm. Variability in gate distance was observed not only between wafers, but also across individual wafers due to spatial variations in the etch rate.

### 4.2.4 L3: Electrode Deposition

The electrodes for the CNT are deposited using a 2 nm chromium adhesion layer, followed by a 100 nm palladium layer. Palladium was selected as the electrode material because it forms a favorable Schottky barrier with CNTs and provides excellent electrical contact properties, as demonstrated in previous studies [135]. In addition to its contact performance, palladium also exhibits high resistance to dry etching, which adds robustness during further processing. Unlike gold, palladium forms a native oxide in ambient conditions, which may require additional surface treatment prior to CNT transfer, as discussed in Section 4.2.9.

As with the gate patterning step, high alignment accuracy is required in the region of the electrodes, while alignment precision in other areas is less critical. The metal layers are deposited using a lift-off process. Since there is no available lift-off resist with dual sensitivity to both UV and electron-beam exposure, the electrode metallization is carried out in two separate steps.

In the first step, the electrode regions are patterned using the same resist stack as in the marker definition (Section 4.2.1), consisting of AR-P617.06 and AR-P672.02. The exposure is carried out by electron-beam lithography as a service at BRNC, followed by development. A 2 nm chromium layer and a 100 nm palladium layer are then deposited by evaporation. Lift-off is performed and then the wafer is stripped.

For the second deposition step, which uses optical lithography, the wafer is first heated on a hotplate for 5 min to dehydrate the surface and improve resist adhesion. The sacrificial layer LOR-5B is then spin-coated, followed by the patterning layer AZ 1512. Just before evaporation, a brief plasma cleaning step is conducted within the vacuum chamber of the evaporator to remove residual surface contamination. This cleaning slightly etches the surface, typically by a few nanometers, but ensures improved metal adhesion. After evaporation, the lift-off is carried out in DMSO at 130 °C for at least 1 h to fully strip the LOR-5B and reveal the final metal structures.

### 4.2.5 L4: Beam-Shape and Actuator Patterning

In this step the device layer if fully patterned to define the beam structures, actuators, and electrodes, which requires a silicon etch. Here, the electrode region again requires a critical alignment. A key challenge is the precise separation of gate electrodes, where the trench width must be minimised to allow for either shorter suspended CNT lengths or wider electrodes. Although deep reactive ion etching (DRIE) is well known for achieving high aspect ratios, reliably etching such narrow and high aspect-ratio trenches remains a technical challenge.

The DRIE process was developed in collaboration with SPTS Technologies GmbH, which provided the Alcatel AMS 200SE I-Speeder tool to the BRNC. To validate the process, test chips were diced from the same SOI wafers used in device fabrication. These wafers consisted of a 500 nm $Si_xN_y$ layer and a 5 µm Si device layer. The test chips were patterned and sent to the company for evaluation. The resist was structured with trench widths ranging from 150 nm to 400 nm to assess the etch capability of the system. The process was considered successful when a 250 nm-wide trench could be etched cleanly using DRIE. Even the trenches with a width of 150 nm were successfully etched, corresponding to an aspect ratio of approximately 37:1. Importantly, the optimized etch exhibited no observable notching at the $SiO_2$ interface, indicating excellent anisotropy and process control.

The proposed process was able to remove both the $Si_xN_y$ isolating layer and Si device layer in one step (Fig. 4.7a). As the BRNC DRIE chamber permits only silicon etching, an additional reactive ion etching (RIE) step was required beforehand to remove the $Si_xN_y$ layer. To withstand the prolonged plasma exposure during the subsequent etch steps, AR-N 7520.17 was spin-coated at 1000 rpm to achieve a thickness of around 800 nm and post-baked at 85 °C for a few minutes. The exposure process was similar to that used for gate patterning, though longer exposure times were employed to ensure adequate resist definition. After RIE and DRIE, the remaining resist was stripped using an oxygen plasma process. The detailed process is found in Appendix C.4.

**(a)** SEM image of a patterned beam structure, showing the layered composition of the actuator: silicon nitride, silicon, and the deposited Cr/Pd metal stack. In the center of the beam, $Si_xN_y$ is removed to reduce stress induced deflection.

**(b)** SEM image of the full MEMS actuator with the central mesa structure etched into the silicon handle layer. The mesa provides space for the CNT growth substrate during dry transfer. Both the mesa plateau and the silicon under the actuators are removed from the backside during the L6 etching step.

**Figure 4.7:** Beam patterning (L4) and mesa etch (L5).

## 4.2.6 L5: Mesa Etch

To prepare the substrate for the CNT dry transfer, space needs to be created for the CNT growth substrate to move along the electrodes. The mesa depth was 35 µm and was etched with DRIE, after the BOX layer was removed using BHF (Fig. 4.7b). To prevent the BHF from attacking the $Si_xN_y$ in the electrode region, special care is taken during the preparation of the photoresist. To ensure maximum adhesion, the wafer was dehydrated at 180 °C for 5 min before the application of a HMDS adhesion layer. Following exposure, the resist was then post-baked. Shock-cooling on the metal heat-sink was avoided to minimize the risk of cracks forming that would allow BHF to reach $Si_xN_y$ and propagate at the interface.

The sample was placed in BHF for 24 min and then carefully rinsed in water. Then the standard DRIE Si etch process was employed for 35 cycles. Stripping was done using DMSO at 130 °C for minimum 1 h.

## 4.2.7 L6: Backside Etch and Release

The LPCVD from step L0 (Section 4.2.1) deposited $Si_xN_y$ on both sides of the wafer. To prepare the backside etch, all of the backside nitride was removed using RIE. To prevent damage to the topside structures during this process, a protective photoresist layer was spin-coated and baked.

For the backside patterning, AZ 4462 positive photoresist was spin-coated at 2000 rpm and baked at 110 °C for a minimum of 10 min. The exposure was performed using a mask aligner with backside alignment capability. After the development, the wafer was placed onto a carrier wafer with wax on a hotplate. The carrier wafer was necessary to prevent etch-through in the process chamber and the exposure of the chuck to plasma.

During the backside etch, the structures become fragile. The carrier wafer also plays an essential role in maintaining structural stability and preventing device break-out or debris from contaminating the chamber. Uniform distribution of the wax layer is especially important, as it serves as a thermal conductor ensuring even heat transfer and preventing local overheating of the exposed wafer. Uneven wax layers or trapped air bubbles can result in the resist to burn locally.

The DRIE backside etch required a minimum of 450 cycles. The process heavily consumes the resist material, necessitating careful optimisation. Achievable aspect ratios for the resist thickness were explored together with Etienne Sciaroni in his bachelor's thesis, yielding a minimum width of the unmasked (open) area of 100 µm for successful etching through the 400 µm thick handle layer, before consuming the photoresist.

Following DRIE, the MEMS structures were released using BHF. During this release step, delamination of the palladium electrodes was observed (Fig. 4.8a). This phenomenon was also reported in earlier experiments involving PECVD-deposited $Si_xN_y$ and even when protective layers were applied using atomic layer deposition of $Si_xN_y$ and $SiO_2$. The root cause of the delamination was traced to the non-negligible etch rate of $Si_xN_y$ in BHF. While this etch rate is typically negligible, it becomes significant when working with narrow electrodes and extended immersion times. After observing this delamination with PECVD deposited silicon nitride, LPCVD $Si_xN_y$ was used instead, which exhibits slower etch rates in BHF.

The required etching time for full release of the MEMS in BHF is around 24 min, but delamination was observed after just 13 min. To further mitigate the risk of delamination, the exposure time of the top side of the wafer to BHF needed to be further reduced. To achieve this, the wafer was not separated from the carrier wafer after DRIE, but directly placed into BHF with the top side still fully covered with the protective photoresist. This allowed the BHF to etch the BOX layer only from the bottom side, without the top side including the electrodes beeing exposed. After 22 min in BHF, the wafer was rinsed and both the wax and the carrier wafer were removed in DMSO at 130 °C. Then, the wafer was further etched in 1 min increments until all $SiO_2$ was etched to prevent delamination due to over etching.

The removal of the BOX layer is shown in Figure 4.8b between device and handle layer.

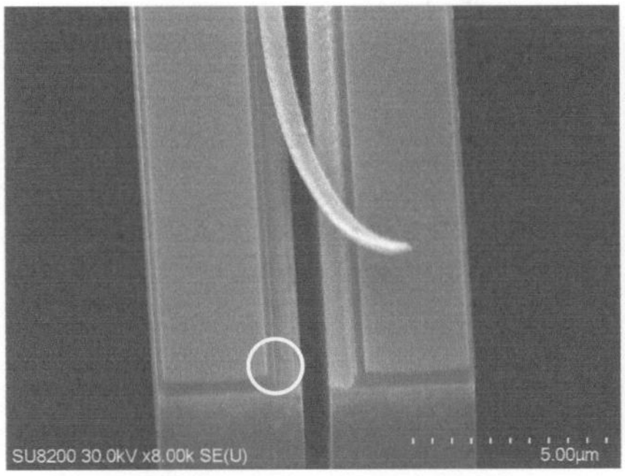


**(a)** SEM image of a fully delaminated gate electrode. Delamination is also observed at the corners of the source and drain electrodes, as highlighted by the circle.

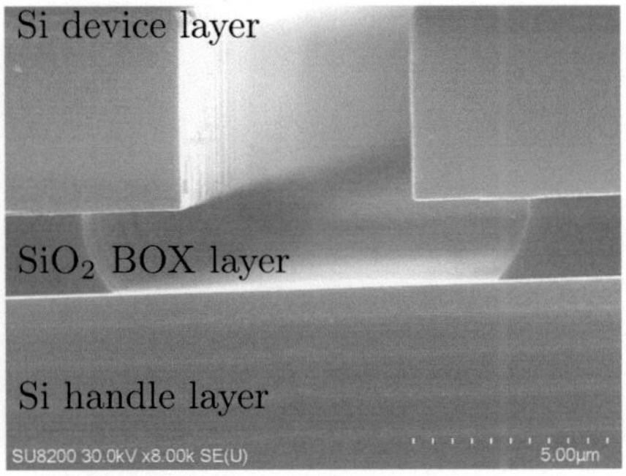


**(b)** SEM image showing the etched BOX layer after the release etch. The suspended Si device layer is clearly visible above the etched cavity, with the remaining BOX layer and Si handle layer beneath.

**Figure 4.8:** BHF release step. The images show structural effects resulting from the wet etch process, including the successful removal of the buried oxide layer and unintended metal delamination.

## 4.2.8 Device Assembly and Wirebonding

The MEMS devices are separated from the wafer using designated tweezer-release-holes. These are designed to allow for the controlled breakage and release of the individual die at the intended breaking points (see Fig. 4.5). The devices are then assembled into a ceramic 32-contact package on a stack of old device dies using conductive silver paste. The stack is required to make space for the growth substrate during the dry-transfer process and prevent it to bump into the sides of the carrier package. The devices are carefully inspected for any damage after this step. If an excessive amount of conductive paste is applied, it may squeeze through the backside etched hole and reach the thermal actuators. The devices are then wire-bonded using a wedge-wedge method.

Before transferring the CNT, the MEMS substrate undergoes a thorough inspection for shorts. The check ensures correct isolation between each electrode and the isolation between all electrodes and the actuator Si layer. Here, faulty actuators can be sorted out. Figure 4.9a shows a schematic of the measurement points. The resistance between point 1 and 2, labelled as $R_{1,2}$, and $R_{3,4}$ includes the resistance of the actuators together with the contact resistance of the wire-bonds. Results of these measurements are shown in Figure 4.9b. In addition to checking the resistances for irregularities, each actuator has visually been checked under an optical microscope if it moves. Resistances from one side to the other of the

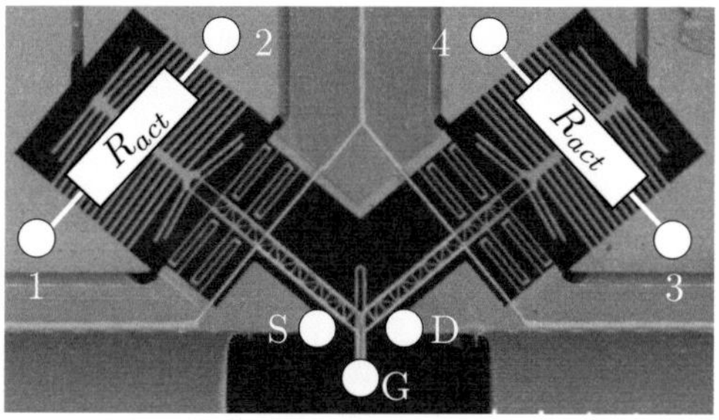


(a) Electrical measurement schematic used to identify properly functioning devices. The resistance of each actuator arm $R_{act}$ is measured between the indicated contact pads.

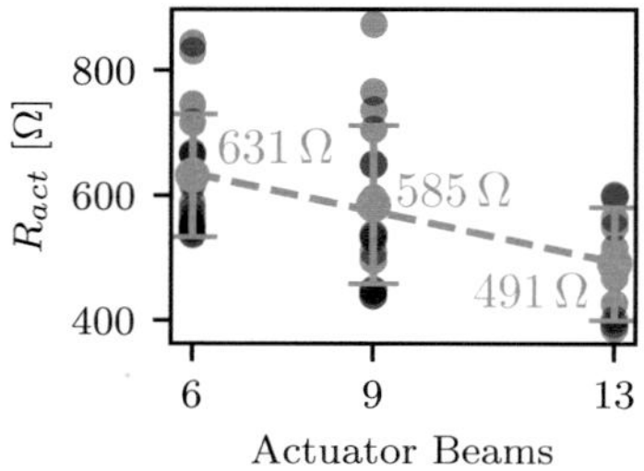


(b) Measured actuator resistances $R_{act}$ for devices with different numbers of actuator beams. A decreasing trend in resistance is observed with increasing beam count.

**Figure 4.9:** Electrical MEMS characterization and actuator resistances. The figure shows the electrical terminals for resistance measurements used to identify working devices.

actuator, e.g. $R_{1,3}$ or $R_{2,4}$ could indicate tiny breaks in the thin connecting beam between actuator and electrodes.

The variance of measured actuator resistances $R_{act}$ is attributed to the varying position and quality of the wire-bonds, while the resistance of the actuators themselves is assumed to be less variant. To avoid the influence of the wire-bonds, the actuators are driven with current.

Source, gate and drain electrode require good isolation for successful measurement. Each device with $R_{S,G}, R_{G,D}, R_{S,D} < 200\,\mathrm{M\Omega}$ was rejected. A good isolation between the silicon device layer, that carries the currents for the actuators and the source, gate and drain electrodes is required. Any resistance $R_{1,S}, R_{2,G}, R_{3,S}, ...$ or other combinations smaller than $100\,\mathrm{M\Omega}$ is excluded.

## 4.2.9 CNT Dry-Transfer

The CNT dry-transfer technique consists of the placement of CNTs, initially grown on a separate substrate resembling the prongs of a fork, onto the MEMS-based straining platform. The growth substrate, which carries the CNTs, is positioned with precision using a micromanipulator above the electrodes and then lowered so that the CNT is placed onto them (see Fig. 4.10a).

The CNTs were synthesised on the growth substrates via a chemical vapour deposition (CVD) process conducted at the BRNC by Dr. Seoho Jung. In

**(a)** CNT growth substrate aligned over (out of focus) the electrodes.

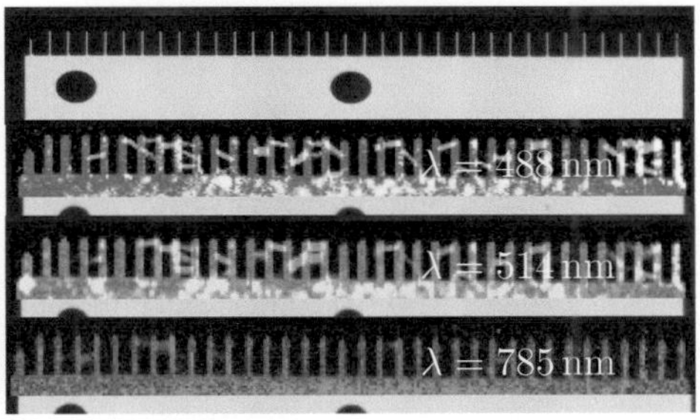


**(b)** CNT growth substrate with Raman maps of different wavelengths λ. Image courtesy Miroslav Haluska.

**Figure 4.10:** CNT transfer procedure and Raman localized CNTs on the growth substrate.

this process, iron-loaded ferritin particles were employed as catalytic agents to facilitate CNT growth [136, 137].

To minimise contact resistance between the CNTs and the metal electrodes, an Argon plasma cleaning step was performed using the sputter tool at BRNC (25 W, 5 min) [92]. This pre-treatment significantly enhanced the electrical contact by removing surface contaminants and the native oxide that formed on the palladium electrodes. Following this, the CNTs were carefully transferred onto the MEMS devices.

Post-transfer, the devices underwent thermal annealing using atomic layer deposition (ALD) at 250 °C for 1 h in vacuum. This drastically improved the contact resistances [92].

Prior to the transfer, Raman spectroscopy was utilised to localize the CNTs on the growth substrate using different wavelengths (see Fig. 4.10b). The localization allowed for excluding trenches with multiple CNTs, entangled CNTs, or non-horizontal CNTs. These characterisation measurements were conducted by Dr. Miroslav Haluska.

## 4.3 Characterization of the Thermal Actuators

To characterize applied strain $\epsilon_A$ or applied tension $T_{ax}$ on a CNT resonator, the applied displacement $\Delta l$ of the actuator needs to be precisely characterized. Because of the vacuum chamber (Section 5.2) where the resonating experiments were conducted, direct measurement of the displacement like Ning et al. performed [70] are not possible in the measurement setup. Available measurement approaches are a white light interferometer operating at room temperature, a vibrometer, or an SEM. As discussed before, the heat dissipation path through air could heavily influence the displacement performance of the actuator which excludes the WLI as option.

The SEM offered similar experimental conditions as the vacuum chamber used for experiments in terms of pressure and temperature.

### 4.3.1 Experimental Setup

The acquisition of most SEM images was part of a student project of Lorenzo Fulciniti.

For SEM image analysis, two Keithley 2400 powersource meters were connected through feedthroughs into the SEM chamber using the same printed circuit board (PCB) and connectors as for the resonance experiments (see Section 5.2). The PCB was carefully mounted onto the SEM holder using the SEM double sided black tapes.

The pictures were taken at a working distance of 6 mm. Pictures were then taken for increments of $I_{act} = 1\,\text{mA}$ up to around $I_{act} = 25\,\text{mA}$. The experimentally applied currents were far beyond the limit of regular operation, which went up to 10 mA current per actuator maximum. At higher currents, Joule heating raises the temperature further, driving silicon into its intrinsic semiconducting regime. As shown in this section, for small currents, the extrinsic region can be well described by a quadratic fit.

Three selected images of such image acquisition are shown in Figure 4.11 of device MV 01 for actuator currents $I_{act}$ of 0 mA, 10 mA, and 21 mA.

### 4.3.2 Displacement Extraction of SEM Images

To extract the current dependent displacement from the SEM images, each image is analysed individually for detectable edges that could be tracked. These edges are mainly the edges of of the Pd-electrodes, but if the SEM

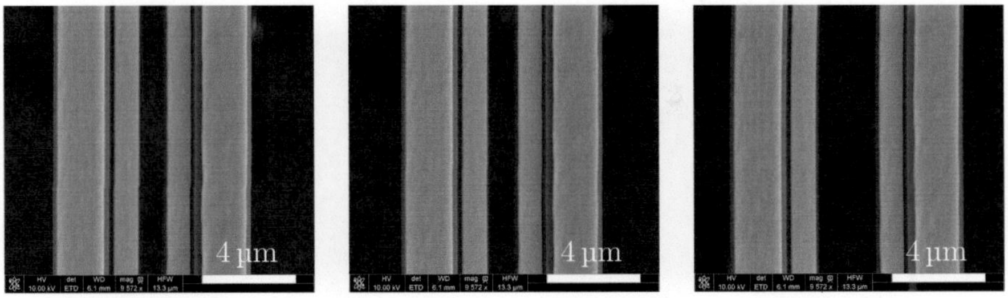


**(a)** $I_{act} = 0\,\text{mA}$. **(b)** $I_{act} = 10\,\text{mA}$. **(c)** $I_{act} = 21\,\text{mA}$.

**Figure 4.11:** SEM images of the separated electrodes, while driving the actuators at different currents. The displacement of electrodes becomes clearly visible for high currentts. Picutres taken by Lorenzo Fulciniti. Device MV 01 (A13).

image quality is high enough can also include the step in $Si_xN_y$ for the gate (visible in Fig. 4.11). For each image, a range of y-coordinates of pixels is given to exclude for example particles that might influence the extraction process.

The common *Laplacian of Gaussian* (LoG) method is applied, that consists of a Gaussian filter that smooths the image and reduce noise, which is necessary for the Laplacian filter that identifies the second derivative of intensity changes. The second derivative of intensity is then averaged of the given interval of y-axis to show a one-dimensional profile line. The edges are located at the roots of the result. The result of one exemplary edge extraction is shown in Figure 4.12.

The x-coordinates are at this point still in pixel and need to be translated into meaningful spacial coordinates in nm. To translate the pixels, the scale bar was used to extract nm per pixel using `ImageJ`. To make this easier, the same working distance and magnification was used for each SEM image for a device measurement series. Since the electrode structure is mirrored at the separating gate gap, the equivalent edges are grouped (same colors in Fig. 4.12) and their x-coordinates are subtracted to get the distances $d_i$ in between them. The amount of detectable edge-pairs varies from sample to sample between 4 and 8.

Once the distances $d_i$ are obtained for every current, the distances are translated into displacements by subtracting the initial distance of each edge-pair at $0\,\text{mA}$ actuator current. This process is susceptible to errors. In order to reduce this error a quadratic function is fitted onto the distances (grouped in a vector $\mathbf{d}$) for low currents (see Section 4.4.2).

$$\mathbf{d} = a \cdot I_{act}^2 + \mathbf{b} \tag{4.2}$$

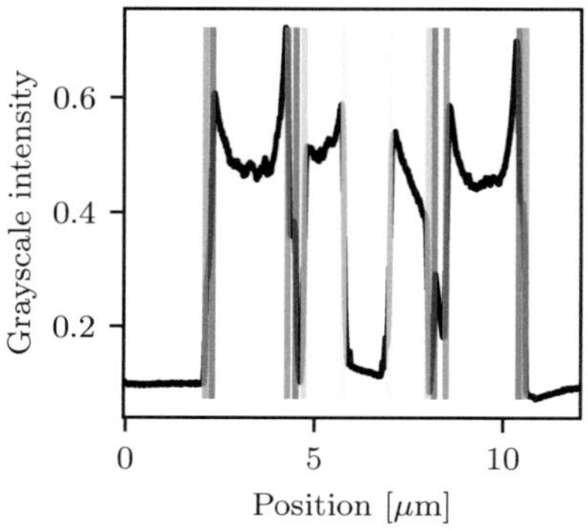


**(a)** Averaged grayscale intensity along profile. Colorpairs indicate edge-pairs as shown in (b).

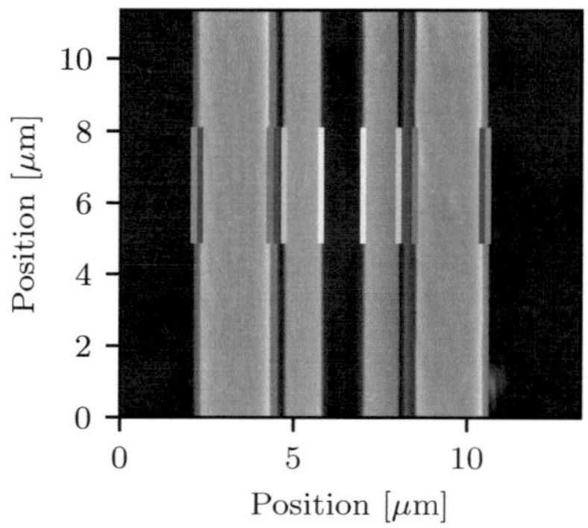


**(b)** SEM image with overlay of extracted edges and extraction range. SEM image taken by Lorenzo Fulciniti.

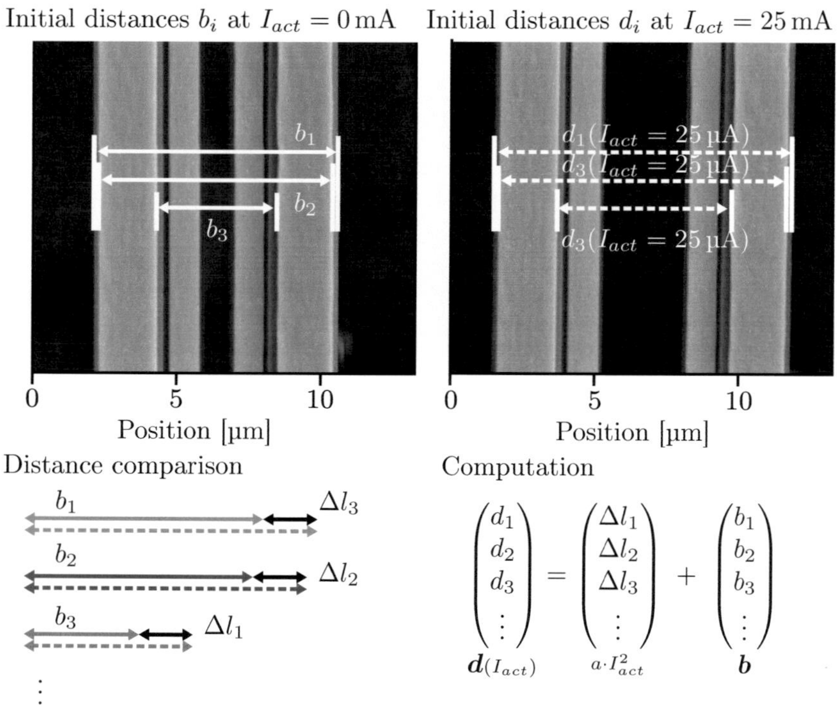


**(c)** Visualization of the extraction method for the actuator current dependent displacements. The initial distances $b_i$ at $0\,\text{mA}$ are subtracted from the distances $d_i$ for each actuator current.

**Figure 4.12:** SEM data extraction with image processing. a) averaged profile line of the gray scale SEM image. The edges have been identified by its slope (derivative). b) overlay with SEM image for verification.

where $a$ is the displacement coefficient and $\mathbf{b}$ is the vector of the initial distances between the edge-pairs. The accuracy here is increased, by forcing the same displacement-coefficient $a$ onto all edge-pairs. An example is shown in Figure 4.13. All other device data extraction are shown in Appendix E.

This process was done for 2 devices with the A6-design, 4 with the A9-design, and 3 devices with the A13-design. All results are plotted in Figure 4.14. The amount of beams in the chevron structure heavily influence the current dependent displacement. The zoom on the right for lower currents show the expected quadratic behaviour according to the joule heating (see Section 4.1.1). For increasing currents however, the relation shows a more rapid increase, followed by a decrease of displacement for even higher currents. This behaviour is due to the semiconducting nature of the Si of which the actuators are made of. This behaviour is modelled and explained in the next Section 4.4.

In Figure 4.14 device L7 is an outlier and displaces much less then the other actuators of its kind. The reason for this behaviour is shown in Figure 4.15. A huge particle is sitting on top of the beams.

The range of the actuator currents for the SEM extraction were limited to protect the actuator from reaching its melting temperature. In an extra experiment an actuator current of 45 mA triggered visible thermal radiation shown in Figure 4.16a. It can be seen that the glow is not exactly symmetrical around the structure, which is most likely due to an alignment inaccuracy of the backside layer etch (shown in Fig. 4.19a), that heavily influences heat dissipation in the substrate. It can also be due small thickness variations from the fabrication process. Figure 4.16b and 4.16c show melted actuator structures once the actuator current reached 25 mA of device MV 20.

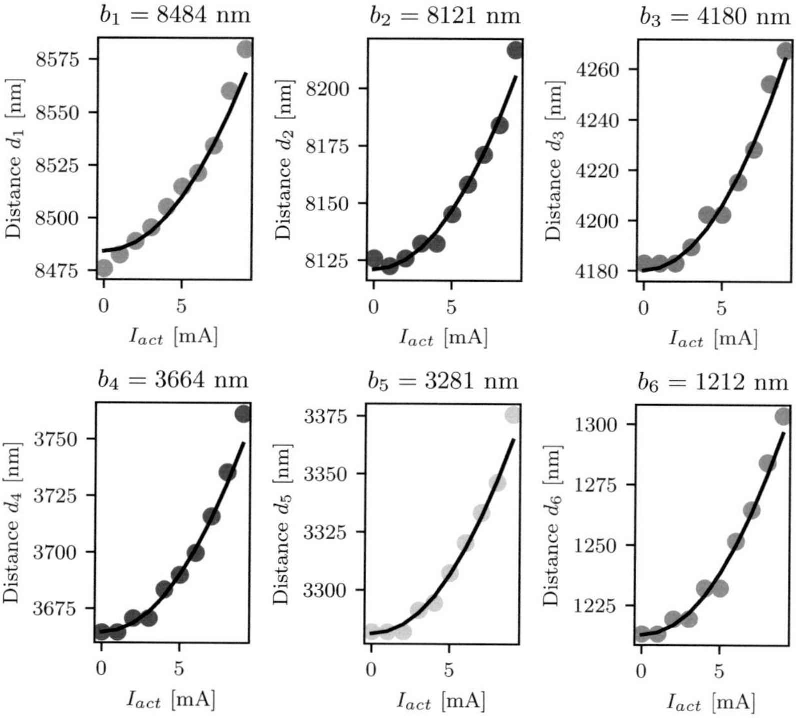


**(a)** Distances $d_i(I_{act})$ measured between each edge-pair $i$ (shown in the same colours as in Figure 4.12), with the initial distance $b_i$ obtained from fitting Equation 4.2 . Each subplot shows the quadratic fit individually to each edge-pair.

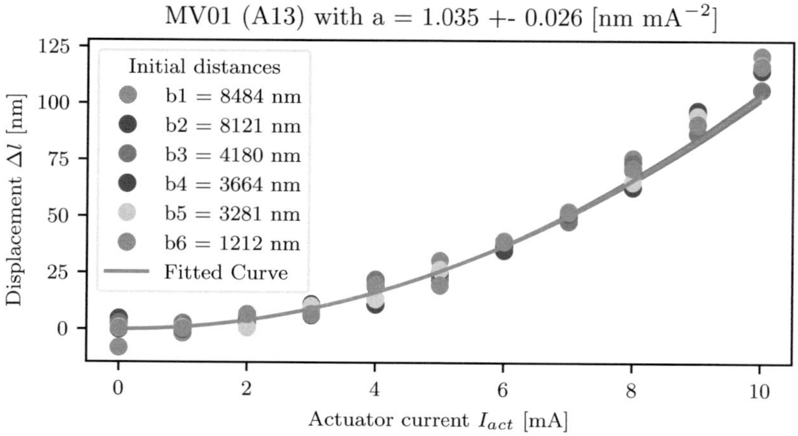


**(b)** Displacement $(d_i(I_{act}) - b_i)$ for each edge-pair, plotted against actuator current $I_{act}$. The combined data shows excellent agreement with a quadratic fit. Here, the fit for MV 01 is shown with $\Delta l = aI_{act}^2$, where $a = (1.035 \pm 0.026)\,\mathrm{nm\,mA^{-2}}$.

**Figure 4.13:** Fitting process of the quadratic function to extract the current-displacement characteristics from the SEM images.

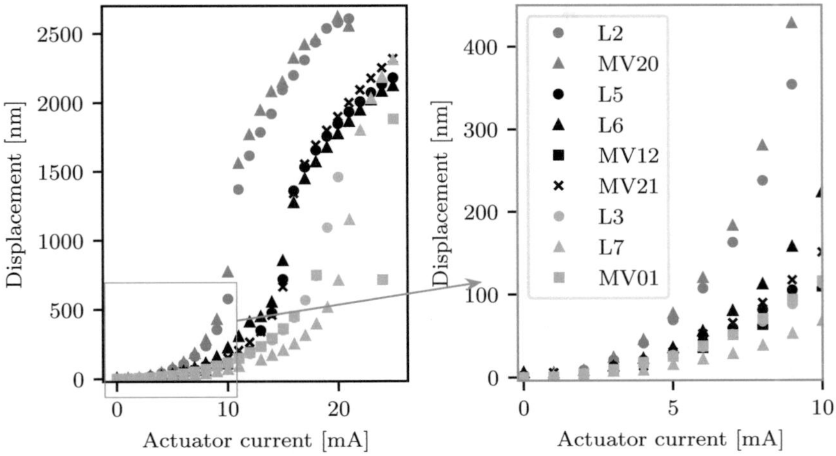


**Figure 4.14:** All characterized MEMS straining platforms separated by the actuator type. Right side shows a zoom on the quadratic region. All devices, their designs and displacement characteristics are shown in Table 4.5.

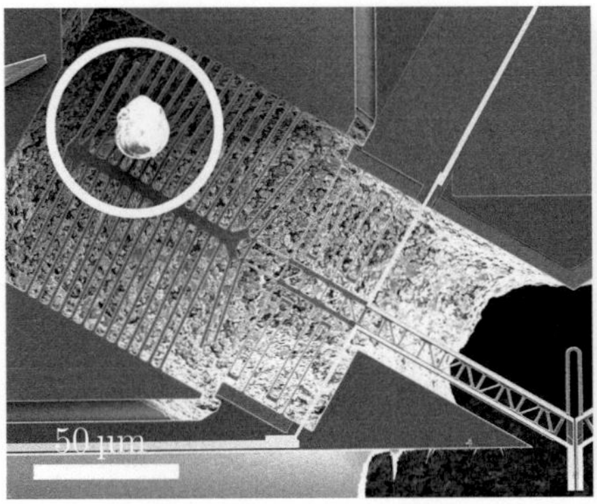


**(a)** SEM image of actuator with attached particle (circled in red).

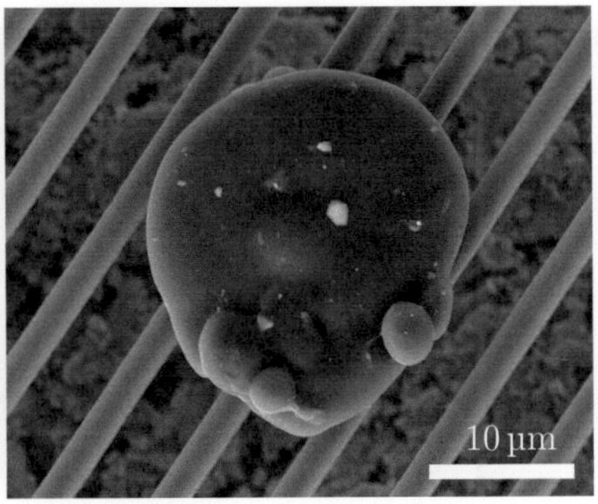


**(b)** Close-up of the particle, possibly originating from fabrication contamination.

**Figure 4.15:** Device L7 showed much lower displacements than actuators of the same design. One actuator of device L7 has a big particle attached to it, reducing its displacement.

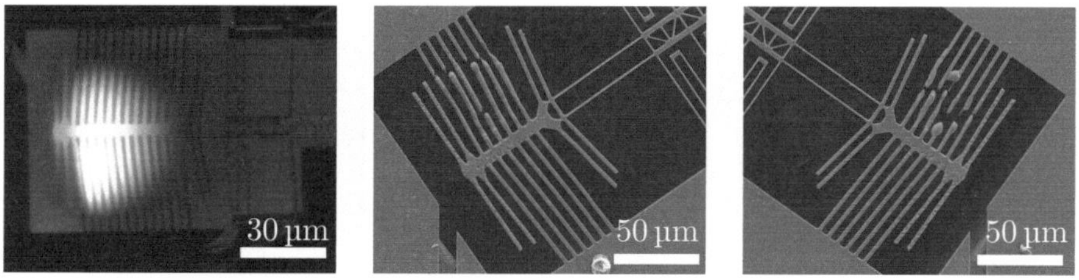


**(a)** Actuator (A13) glowing at ambient pressure and $I_{act} = 45\,\text{mA}$.
**(b)** Burned actuator MV20 (left).
**(c)** Burned actuator MV20 (right).

**Figure 4.16:** Limitations of the thermal actuators. While actuators with 13 beams still function and glow at high currents of 45 mA (ambient pressure), actuators with less beams (A6) melt already at 22 mA (observed in vacuum).

# 4.4 COMSOL FEM Simulation of the MEMS Straining Platform

The previously shown measurements have shown quadratic displacement behaviour for low currents, but for higher currents the quadratic relation does not hold true any more. To investigate and further understand this behaviour, a FEM study is employed using COMSOL.

## 4.4.1 COMSOL Model

The actuators are made out of highly doped Si, which has semiconducting properties, therefore its electrical conductivity is highly temperature dependent, because more charge carrier get available for higher temperatures. To capture this complex temperature dependent behaviour, all material properties such as the charge carrier concentration $n_i(\mathrm{T})$, thermal conductivity $\kappa(\mathrm{T})$, thermal expansion coefficient $\alpha(\mathrm{T})$ and electrical conductivity $\sigma(\mathrm{T})$ were carefully modelled according to literature.

The simulation is performed in 2D the modules *solid mechanics*, *heat transfer*, and *electric currents* with the multiphysics modules *electromagnetic heating* and *thermal expansion* to cover codependent relations. Figure 4.17 the interacting and dependent relation between all COMSOL modules.

The semiconducting electrical conductivity $\sigma(\mathrm{T})$ is textbook derived [138, 139] and is described in dependence of the electrons with mobility $\mu_e(\mathrm{T})$

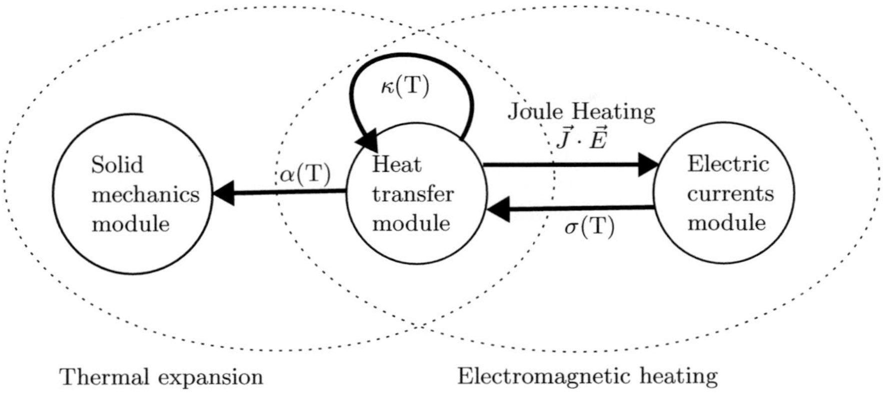


**Figure 4.17:** Schematic of the different simulation and Multiphysics modules in COMSOL.

and their intrinsic carrier concentration $n_i(\mathrm{T})$ and the holes with mobility $\mu_h(\mathrm{T})$. $e$ is the electrical charge.

$$\sigma(\mathrm{T}) = n_i(\mathrm{T}) \cdot e \cdot \mu_e(\mathrm{T}) + (n_i(\mathrm{T}) + p_{dop}) \cdot e \cdot \mu_h(\mathrm{T}) \tag{4.3}$$

The intrinsic carrier concentration $n_i(\mathrm{T})$ is temperature dependent [140] and has been characterized as:

$$n_i(\mathrm{T}) = 5.71 \times 10^{19} \cdot (\mathrm{T}/300)^{2.365} \cdot \exp(-6733/\mathrm{T}) \tag{4.4}$$

The temperature dependent charge carrier mobility in silicon has also been investigated in literature [141]:

$$3\mu_h(\mathrm{T}) = \mu_e(\mathrm{T}) = 1.5 \times 10^6 \cdot \mathrm{T}^{-3/2} \left[K^{3/2} cm^2/V/s\right] \tag{4.5}$$

The exact doping concentration $p_{dop}$ of the used SOI wafer device layer was not specified by the manufacturer and is used as fitting parameter to match the transition point from intrinsic to extrinsic region of the semiconductor. A lower bound could still be estimated using [139, p. 30]:

$$\rho_{SI} = \frac{1}{q\mu_p p_{dop}} < 0.02\,\Omega\,\mathrm{cm} \tag{4.6}$$

$$p_{dop} > 3.1 \times 10^{18}\,\mathrm{cm}^{-3} \tag{4.7}$$

**Table 4.3:** Summary of simulation parameters for the silicon actuator.

| Silicon material simulation parameter | Value | Comment |
|---|---|---|
| Ambient temperature $T_{ambient}$ | 293.15 K | Lab environment |
| Device layer thickness $d_{Si}$ | $(5.0 \pm 0.5)$ µm | Manufacturer |
| Device layer resistivity $\rho$ | <0.02 Ω cm | Manufacturer |
| Density $\rho$ | 2330 kg m$^{-3}$ | COMSOL |
| Relative permittivity $\epsilon_r$ | 11.7 | COMSOL |
| Anisotropic elasticity Matrix | see F.1 | COMSOL |
| Heat capacitance $C_p$ | 704.2 J kg$^{-1}$ K | COMSOL |
| Thermal expansion coefficient $\alpha$ | See Eqn. 4.8 | From [143] |
| Electrical conductivity $\sigma$ | See Eqn. 4.3 | From [138] |
| Intrinsic carrier concentration $n_i$ | See Eqn. 4.4 | From [140] |
| Charge carrier mobilitity $\mu_h, \mu_e$ | See Eqn. 4.5 | From [141] |
| Thermal conductivity $\kappa$ | See Fig. F.1 | From [142] |
| Heat transfer coefficient $h_d$ | $\frac{\kappa(\mathrm{T})}{400\,\mu\mathrm{m}} \cdot (h_0 + h_1 \cdot \mathrm{T})$ | fitted |

with elementary charge $q = 1.602 \times 10^{-19}\,\mathrm{C}$ and hole mobility $\mu_p \approx 100\,\mathrm{cm}^2\,\mathrm{V}^{-1}\,\mathrm{s}^{-1}$.

The thermal conductivity $\kappa$ is adapted from the work presented in [142], where it was measured every 50 K from 100 K to 1350 K. The values are interpolated in COMSOL with a piece-wise cubic method for the final function (Fig. F.1). The thermal expansion coefficient $\alpha$ used is as provided from [143]:

$$\alpha(\mathrm{T}) = (-15.25 + 3.43 \cdot \ln(\mathrm{T}) - 5.39 \times 10^{-3} \cdot \mathrm{T} + 1.3 \times 10^{-6} \cdot \mathrm{T}^2 - 88.69 \cdot \mathrm{T}^{-1}) \times 10^{-6}, \tag{4.8}$$

where T is the temperature of the material. This concludes the input temperature dependent material properties.

The simulation domain is created from the mask-files for fabrication (Fig. 4.19b). The wire bonds are represented as circles with a 50 µm diameter. The current for simulation is applied on the edges of one circle, the other circle is grounded. The device layer thickness is specified by the manufacturer as $(5.0 \pm 0.5)$ µm (see Appendix C.8). The edges of the suspended regions are mechanically fixed, all supported regions are assigned an out of plane heat flux through the BOX layer into the Si handle layer. This value is unknown and serves as a fitting parameter. A summary of all material parameters is shown in Table 4.3.

Lastly, the out of plane heatflux and the doping concentrations were left to

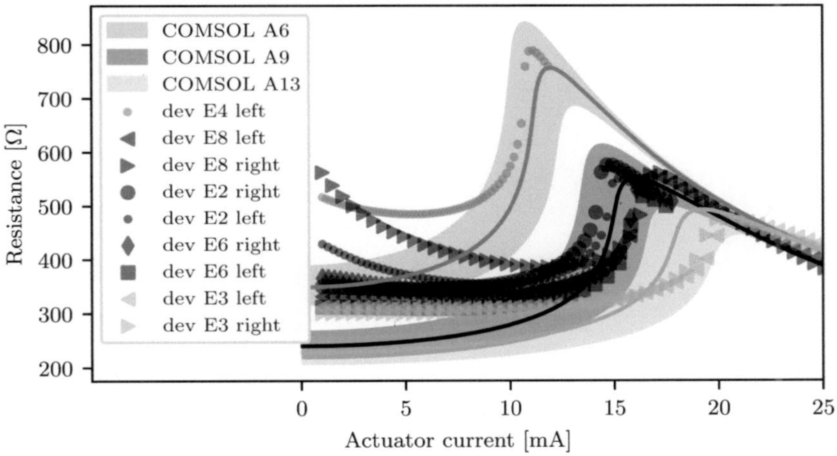


**Figure 4.18:** Actuator resistances $R_{act}$ of different MEMS designs at different actuator current $I_{act}$ measured in vacuum condition $< 1 \times 10^{-3}$ mbar. The measured datasets where corrected by contact resistance values as fit. Important for a good fit is only the tail at high currents and the onset of resistance increase. Resistances were derived from measured voltages across the actuator by an applied current with a Keithley 2400 power source.

be determined. However no reference values were found for either. To match the doping concentration, current sweeps have been performed on chips together with a bachelor student Eric Meier. Then, the doping concentration was fitted to obtain good agreement (see Fig. 4.18). The fitted doping concentration was set to $p_{doping} = 4.5 \times 10^{18}\,\mathrm{cm}^{-3}$, which is a common value for "heavily doped" silicon and in agreement with the estimation in Equation 4.7. The out-of-plane heat flux into the handle layer through the BOX layer is then used to fit the current-displacement curve of the simulation onto the extracted data for the A9 designs. Since the 2D system is missing a third dimension, the out of plane heatflux was modelled as effective conduction per thickness as:

$$h_d(\mathrm{T}) = \frac{\kappa(\mathrm{T})}{400\,\mu\mathrm{m}} \cdot (h_0 + h_1 \cdot \mathrm{T}) \tag{4.9}$$

The fitting parameters have been determined to be $h_0 = 5 \times 10^{-7}$ and $h_1 = 1 \times 10^{-7}\,\mathrm{K}^{-1}$.

Figure 4.19a shows an optical microscope image of a device from underneath. It is visible that the backside etch is not perfectly symmetrical which will

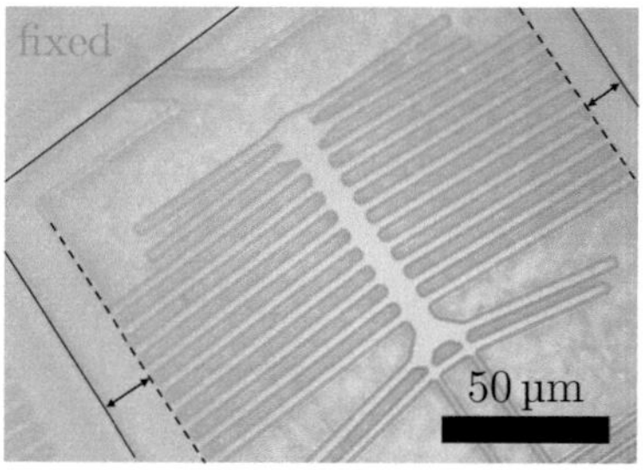


**(a)** Image from the backside of the wafer focused onto to bottom side of the chevron structure. Misalignments can influence the symmetry of the temperature profile in the thermal actuator.

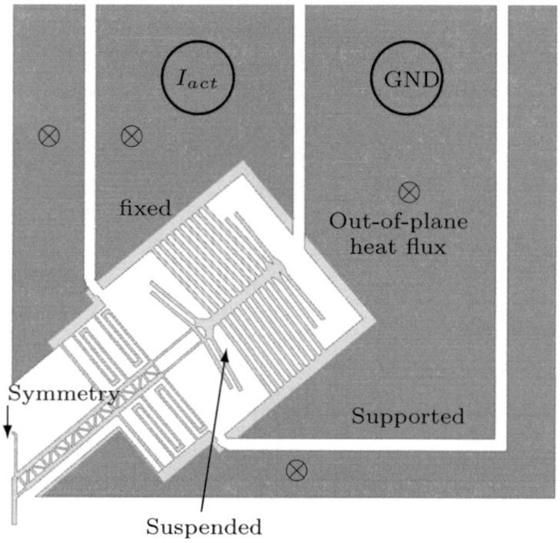


**(b)** The geometry used in the COMSOL simulation with the current terminal $I_{act}$ and the symmetry point with .

**Figure 4.19:** COMSOL simulation domains and potential inaccuracies for simulation

influence the temperature profile of the actuator. It explains the asymmetric glowing in Figure 4.16a.

The simulations were performed on the ETH Euler Cluster and includes varying thicknesses of the silicon device layer according to the manufacturer specification range between 4.5 µm and 5.5 µm (Appendix C.8).

### 4.4.2 Simulation Results and Analysis

The simulation results are shown together with the SEM extracted data points in Figure 4.20. The simulation includes the specified thickness variation of the silicon device layer by the manufacturer and is represented in the coloured area.

The result shows three different regions for each design. For very low currents, the current displacement relation is quadratic as it follows the joule heating relation $I^2R$. In that region, the temperature increase is so small, that the resistance value does not change significantly. Once the current increases further, this relation is no longer valid as the resistance is temperature dependent. The region in which the quadratic approximation is valid is called *quadratic region* and is further investigated, as this allows the precise description of the actuator displacement and limits the operating range, which can be approximated with a quadratic function.

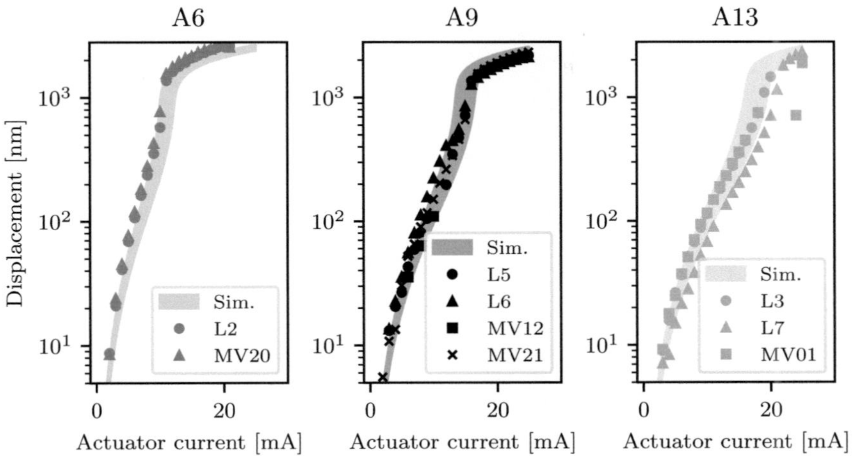


**Figure 4.20:** Simulation results from COMSOL and data from SEM data extraction. The simulation results reflect the device layer thickness uncertainty specified by the wafer manufacturer. All devices show good agreement with the simulation results over a large range of actuator currents.

If the current and therefore the temperature further increase, the semiconductor is still in the extrinsic regime, meaning that its resistance is dominated by electron scattering. Other than for the quadratic region, here the resistance increases significantly with temperature and the displacement increases faster than quadratically.

At some point the semiconductor then reaches its intrinsic region, where the resistance is reduced with larger temperature, as more charge carries become available. This results in return in a decrease of the slope of the displacement.

Both transition points are analysed and defined as follows: The transition out of the quadratic regime happens once a quadratic fit scores a $R^2$-values of less than 0.999. This value is chosen very high indicating and excellent fit of the quadratic function. The transition point into the intrinsic region is defined as where the slope starts decreasing again and is identified by the root of the second derivative. The regions are shown in Figure 4.21.

The results of the region analysis are summarized in Table 4.4. The errors represent the different device layer thickness. With knowledge about the quadratic region, it is now possible to fit Equation 4.2 onto the extracted displacements from Section 4.3.2.

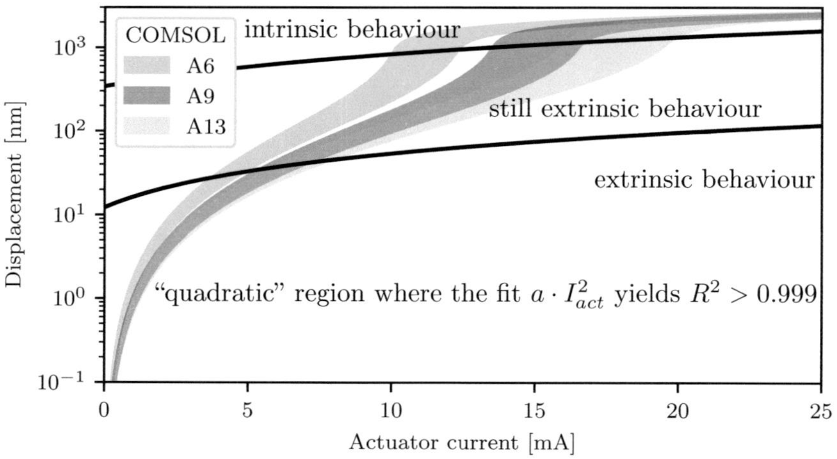


**Figure 4.21:** Regions of different dominant physics regimes for the thermal actuators. For low currents, the displacement follows a quadratic relationship, because changes in resistance remain small, as temperatures remain small. For higher currents, the electron scattering increases and the quadratic behaviour does not hold true any more. For even higher currents, more charge carrier become available decreasing the resistance (intrinsic regime).

**Table 4.4:** Summary of the transition from extrinsic to intrinsic behaviour and the maximum current to obtain a good quadratic fit ($R^2 > 0.999$) for different actuator designs simulations. The errors origin from different simulated thicknesses.

| | A6 | A9 | A13 |
|---|---|---|---|
| Quadratic fit up to [mA] | $4.1 \pm 0.3$ | $5.5 \pm 0.4$ | $6.6 \pm 0.5$ |
| Intrinsic onset current [mA] | $11.1 \pm 0.8$ | $15.0 \pm 1.1$ | $17.9 \pm 1.3$ |
| Quadratic fit up to curr. p. beam [mA] | $0.68 \pm 0.05$ | $0.61 \pm 0.04$ | $0.50 \pm 0.04$ |
| Intrinsic onset curr. p. beam [mA] | $1.86 \pm 0.13$ | $1.66 \pm 0.12$ | $1.38 \pm 0.10$ |
| Displacement coefficient $a$ in the quadratic region [nm mA$^{-2}$] | $1.60 \pm 0.23$ | $0.94 \pm 0.13$ | $0.84 \pm 0.12$ |

## 4.5 Results of the Actuator Characterization

With the information about the current-displacement characteristics measured from SEM and the COMSOL simulation yielding valid ranges for a quadratic fit, all available devices were fitted with a quadratic function (Eqn. 4.2) to extract their displacement coefficient $a$. The determination of the the "quadratic region" was important, because the outcome of the fit showed high sensitivity to the maximum current used for the fitting range. The results of the fits are summarized in Table 4.5.

Four devices of the same architecture (A9, 2 springs, 40° incline) were available to compare directly with each other. The displacement coefficients showed a large variance, where the largest displacement coefficient was 35 % larger than the smallest. In comparison to the simulation, two devices lie within the $1 - \sigma$ range of the error. The displacement coefficient of device L5 ($1.13\,\mathrm{nm\,mA^{-2}}$) lies within expected ranges for the simulated thickness, as for a device layer thickness of 4.5 µm, the simulation predicts a displacement coefficient of $1.15\,\mathrm{nm\,mA^{-2}}$. The last device (L6) of this design is a clear outlier. Its behaviour can not be explained with device layer thickness variations.

With this variance in mind, the influence of the inclination angle of the beam is investigated. The measured devices in the A13 design all had different inclination angles, but the results do not show the expected trend. Lower inclination angles were expected to show a larger displacement coefficient and vice versa. Here, the angles 35° and 40° showed the same results, while 30° showed the lowest displacement coefficient. All fits lie within the expected range. Both devices of the A6 show much larger displacement coefficients than expected.

The design and fabrication was successful to deliver working devices with different current-displacement curves depending on the actuator design. The simulation approach was successful in replicating the actuator behaviour over a large temperature range including the semiconducting property of silicon even for large temperatures in the intrinsic regime. The large variance of the devices however, requires an individual characterisation of each device.

**Table 4.5:** Fitted displacement coefficients $a$ from SEM image analysis of all devices in comparison to simulated displacement coefficients. The uncertainty in the COMSOL simulation $a$ reflects the uncertainty of the device layer thickness.

| Design | Device | Angle | Springs | SEM $a$ | COMSOL $a$ |
|---|---|---|---|---|---|
| A6 | L2 | 30 | 0 | $2.3\,\mathrm{nm\,mA^{-2}}$ | $(1.60 \pm 0.23)\,\mathrm{nm\,mA^{-2}}$ |
| | MV20 | 35 | 2 | $2.61\,\mathrm{nm\,mA^{-2}}$ | |
| A9 | L5 | 40 | 2 | $1.13\,\mathrm{nm\,mA^{-2}}$ | $(0.95 \pm 0.13)\,\mathrm{nm\,mA^{-2}}$ |
| | L6 | 40 | 2 | $1.31\,\mathrm{nm\,mA^{-2}}$ | |
| | MV12 | 40 | 2 | $0.98\,\mathrm{nm\,mA^{-2}}$ | |
| | MV21 | 40 | 2 | $1.01\,\mathrm{nm\,mA^{-2}}$ | |
| A13 | L3 | 35 | 2 | $1.03\,\mathrm{nm\,mA^{-2}}$ | $(0.85 \pm 0.12)\,\mathrm{nm\,mA^{-2}}$ |
| | MV01 | 40 | 2 | $1.04\,\mathrm{nm\,mA^{-2}}$ | |
| | L7 | 30 | 2 | $0.58\,\mathrm{nm\,mA^{-2}}$ | |

# 5 Electrical Conduction in CNT Resonators

The electrical read-out of CNT resonators is more challenging than that of other resonator systems. Graphene resonators, for example, possess larger capacitances, enabling stronger electrostatic actuation and more efficient signal detection. Similarly, membrane resonators are typically compatible with standard measurement techniques such as optical interferometry or network analysers. In contrast, these approaches are not well-suited for CNT devices due to their extremely small capacitance which is significantly lower than the parasitic capacitances present in the system.

This chapter focuses on the transduction mechanism between the mechanical motion of the CNT and the resulting conductance modulation. This modulation enables the detection of the CNT's motion through electrical current measurements. The physical origins of conductance modulation are discussed in detail, and the associated measurement setup is described. Initial results are presented, along with methods for extracting key parameters such as the eigenfrequency and quality factor.

Building on the foundations established in previous chapters, this section considers the influence of static and dynamic oscillation amplitudes on current response. These amplitudes are essential for accurately computing motion-induced currents in the down-mixing measurement technique.

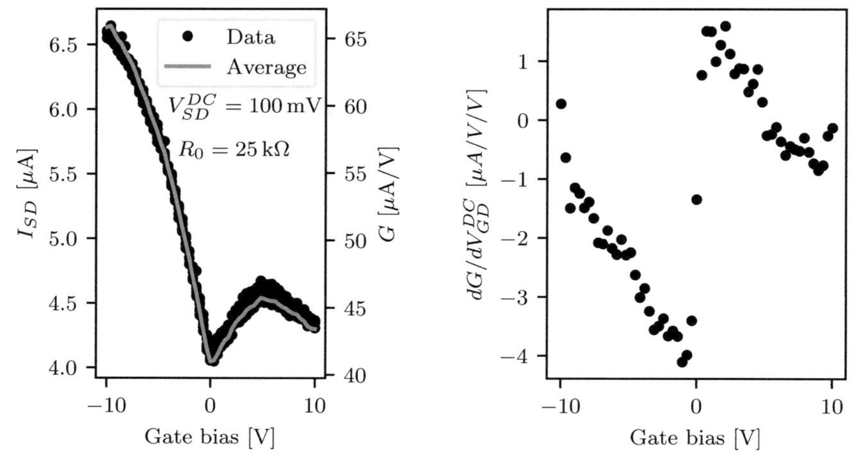


**(a)** Measured and averaged transfer characteristic $I_{SD}$ with applied voltage $V_{SD}^{DC} = 100\,\text{mV}$ across the channel of device CGDS21 shows an SGS CNT behaviour. At $V_{GD}^{DC} = 1\,\text{V}$ the resistance is $R_0 = 25\,\text{k}\Omega$.

**(b)** The derivative of the conductance $G$ with respect to $V_{GD}^{DC}$ represents the modulation of conductance with modulated gate bias.

**Figure 5.1:** Transfer characteristics and differential conductance of a single SGS-type CNT.

## 5.1 Transfer Characteristics of CNT FETs

A CNT resonator resembles a field effect transistor (FET) with the CNT acting as the channel material. Depending on the chirality vector (Section 3.5) CNTs can exhibit different electronic properties: metallic, small bandgap and small bandgap semiconducting (SGS). Figure 5.1a shows the transfer characteristics $I_{SD} - V_{GD}$ of a device (CGDS 21) measured at constant source-drain bias $V_{SD}^{DC} = 100\,\text{mV}$. Measuring the transfer characteristics is essential for interpreting current behaviour during frequency sweep experiments and is conducted both before and after each frequency sweep.

As the measured source-drain current $I_{SD}$ depends on the applied bias $V_{SD}^{DC}$, a more appropriate quantity for CNTs is the conductance $G = \frac{I_{SD}}{V_{SD}}$, which normalises the current to the applied bias. The derivative of the conductance $G$ with respect to $V_{GD}^{DC}$ is called conductance modulation $\frac{dG}{dV_{GD}^{DC}}$.

To characterise the quality of a CNT device, the threshold voltage $V_{th}$ or the on resistance $R_{on}$ can be used. In accordance to Kumar et al. [66], here we use $R_0$ for quantification. $R_0$ is defined as the resistance at 0 V applied gate bias $V_{GD}^{DC}$ and can reach values down to below $100\,\mathrm{k\Omega}$, limited by the channel and quantum resistance. $R_0$ can be extracted from the transfer-characteristics (Fig. 5.1a). This value is important to tunable CNT resonators as it is required to estimate gauge factors (shown in Section 5.3.2.

The source drain current $I_{SD}$ depends on multiple effects. In the channel, the bandgap physics determines the conductivity, which is strongly dependent on the applied gate bias $V_{GD}^{DC}$ as seen in Figure 5.1a, as the gate modulates the carrier concentration in the channel. The effect of the the gate bias modulation onto the conductance modulation can be quantified with the derivative of $G$ with respect to the gate bias $V_{GD}^{DC}$ as shown in Figure 5.1b.

Because of their crystalline structure, some CNTs can also exhibit piezoresistive behaviour, where applied strain can increase or decrease the overall CNT conductivity. SGS CNTs exhibit the highest gauge factors ($GF$), which can exceed 1000 [144]. During resonance, the induced strain by the oscillation modulates therefore the conductivity as well.

Both modulation effects have impact on the over all conductivity and the measured current during resonance measurements. Even though their nature is of different kind, both rely either directly on the modulation of the gate bias ($V_{GD}^{AC}$) or the motion induced strain, that in return depends on the modulation of the gate bias again. This modulation effect is used for the measurement technique explained in the next section. Capacitive displacement currents are not considered due to too small capacitances.

Preferred CNTs for measurements show a very high conductance modulation $g_m$ with modulated gate bias. Metallic CNTs are generally unsuitable for resonance measurements, as they exhibit negligible or no conductance modulation and typically show low gauge factors. Another aspect to keep in mind are the contact resistances, which the CNT forms with the electrodes as well as the leads (cables, wire-bonds etc.). High contact resistances reduce the signal-to-noise ratio and decrease the quality of the measurements. The contact resistances and how to improve them was discussed in the fabrication Section 4.2.9.

The transfer characteristic of CGDS 21 (Fig. 5.1a) has a resistance of $R_0 = 25\,\mathrm{k\Omega}$ , which is unusually low (compare Section 2.3.3). This device had been transferred with 6 different CNTs, which all could decrease the device resistance resistance further in a parallel resistances configuration (see Fig. D.2). However, for CNT resonance measurements, some of them might not be active, making the modelling of the currents more difficult, as it is unclear which CNT is resonating, and which one might be mechanically inactive. A good CNT resonator, should only carry one CNT.

## 5.2 Experimental Setup

An experimental setup (Fig. 5.2) is required to supply static gate and source voltages to measure the transfer characteristics of each device as shown in Section 5.1. The static source voltage was provided by a Keithley 2400 source meter, the static gate bias by a Keysight B2961A with Keysight N1294A noise filter attached to reduce noise influence on the conductance modulation.

At the core of the measurement-technique lies the lock-in amplifier. The measurements on the two devices without axial straining capability (SJDS 123 and CGDS 21, see Appendix D) used a SRS830 lock-in amplifier (the same configuration as [66]). Later, a new lockin-amplifer was introduced (Zurich instrument UHFLI), replacing the old lock-in amplifier and introducing minor changes to the setup, without changing the schematic.

The first setup used two Agilient N5181A signal generators together with the SRS830 lock-in amplifier. The AC read-out signals were split using a power splitter. Then, one side was mixed with a ZX05-10L-S+, amplified in a SRS560 pre-amplifier, which also contained a low-pass filter with a cut-off frequency set at $\Delta\omega$. The pre-amplified signal was then used as a reference signal for the lock-in amplifier. For the $2\omega$ read-out, a frequency doubler FK-5-S was used to double the gate frequency before mixing for the reference signal.

Later, the new lockin-amplifer was introduced. The Zurich instrument UHFLI replaced the SRS830 lock-in and both AC sources, as the new lock-in could generate the AC signal itself. The artificial mixing became obsolete, as the UHFLI was able to mix the frequencies internally, simplifying the whole setup.

In both configurations, the AC and DC voltages for each electrode were added with a bias-tee (ZFBT). To translate the output current at the drain electrode into a voltage for the lock-in amplifier, a trans impedance amplifier (TIA) was used from FEMTO (DHPCA-100). The gain was set to 1000 and kept constant for all measurements.

Source, gate, and drain are fed into a vacuum chamber (lakeshore CPX), connected to a Lakeshore 336 temperature control stage. In this work, only measurements at room temperatures were performed. The vacuum chamber was shielded with a radiation shield inside to reduce further noise. The vacuum was generated by an Agilent pump and pressures were kept below $<10^{-3}$ mbar for all resonance experiments measured by a pressure gauge. The fed through signals were connected via SMA connectors to an impedance matched printed circuit board (PCB) with a socket for the ceramic carrier package with the wire-bonded device.

The communication with the source-meters was through a GPIB protocol and was computer controlled in a python API, which provided pre-programmed experimental procedures such as "transfer-curve", "frequency sweep", or "Gate bias and frequency sweep" with the option to change each parameter. The lock-in amplifier from Zurich instruments used a USB-communication. The lock-in API allows for easy control via python. The pump was controlled via an RS-232 communication protocol. Each data file contains all relevant settings such as the used time-constant.

## 5.3 Frequency Down-Mixing

As seen in the transfer-characteristics in Figure 5.1, the conductance $G$ of the CNT changes with the applied gate voltage. This property is used for conductance modulation in a measurement technique called *frequency down-mixing* which is common for the readout of CNT resonators [55, 145, 146].

The actuation of the CNT resonator is applied through the gate voltage as described in Section 3.3.2 in the form of:

$$\text{actuation:} \quad V_{GD}(t) = V_{GD}^{DC} + V_{GD}^{AC} \cdot e^{i\omega t} \tag{5.1}$$

The Euler representation of oscillation is used in this chapter, as it simplifies the notation drastically.

To measure the CNT current, a voltage signal $V_{SD}(t)$, called read-out signal, is applied between the source and drain electrode. The read-out signal has a small frequency offset $\Delta\omega$ with respect to the actuation signal. Both frequencies will mix in the CNT resulting in a mixed signal that contains a frequency component exactly at $\Delta\omega$. The frequency shift $\Delta\omega$ is typically selected within the kHz range, enabling this component to pass through the low-pass filter characteristics of the overall measurement setup, which includes not only the device itself but also all peripheral components. In this thesis $\Delta\omega = 10\,\text{kHz}$ is valid for all reported data.

Two read-out techniques have been established called $1\omega$ and $2\omega$ read-out:

$$1\omega \text{ readout:} \quad V_{SD}(t) = V_{SD}^{DC} + V_{SD}^{AC} \cdot e^{i(1\omega t + \Delta\omega t + \psi_1)} \tag{5.2}$$

$$2\omega \text{ readout:} \quad V_{SD}(t) = V_{SD}^{DC} + V_{SD}^{AC} \cdot e^{i(2\omega t + \Delta\omega t + \psi_2)} \tag{5.3}$$

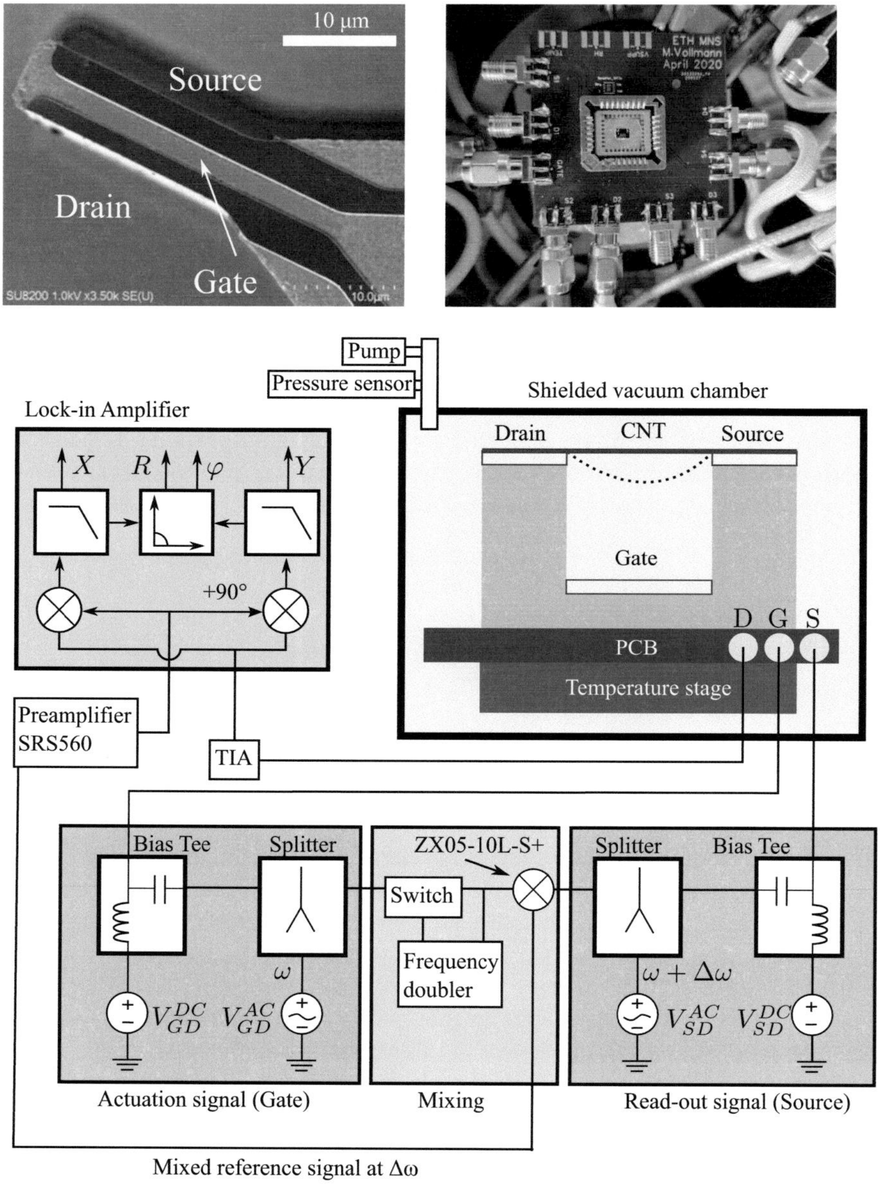


**Figure 5.2:** Experimental setup schematic.

Both read-out techniques yield different contributions in the $\Delta\omega$ component. Two different current contributions and their effective outputs depending on the read-out technique are described in the following sections.

### 5.3.1 Conductive Current

For semiconducting and small bandgap semiconducting CNTs, the conductance $G$ can be modulated by changing the electrostatic gate bias $V_{GD}$. The conductance modulation directly determines the source-drain current. A small charge variation $\delta q$ leads to a conductance modulation, based on the sensitivity of the conductance $G$ to charge changes. This behaviour is approximated in a first order Taylor expansion:

$$G(q + \delta q) \approx G(q) + \frac{\partial G}{\partial q}\delta q \tag{5.4}$$

The sensitivity of the conductance $G$ to charge variation $\frac{\partial G}{\partial q}$ can be extracted from the measured static response shown in Figure 5.1 as the derivative:

$$\frac{\partial G}{\partial q} = \frac{\mathrm{d}G}{\mathrm{d}V_{GD} \cdot C_0} \tag{5.5}$$

where the capacitance $C_0$ has been given in Equation 3.55. The modulated current of this type of charge modulation is called *conductance modulation current* $i_g$ and is written as:

$$i_g(t) = \frac{dG^{DC}}{dV_{GD}^{DC}}\frac{1}{C_0}V_{SD}(t) \cdot \delta q \tag{5.6}$$

The charge modulation $\delta q$ results from the gate voltage modulation over the capacitance between CNT and gate electrode. The conductance is modulated by both the capacitance modulation (change of distance due to CNT movement) and the gate bias modulation:

$$\delta q = \delta(C_g(z,t)V_{GD}(t)) = C_0 \cdot \delta V_{GD}(t) + \delta C_g(z,t) \cdot V_{GD}^{DC} \tag{5.7}$$

$$\delta q = C_0 V_{GD}^{AC} \cdot e^{i\omega t} + V_{GD}^{DC}\frac{dC_g}{dz} \cdot z_d(t) \tag{5.8}$$

The derivative of the capacitance can be understood from the Taylor expansion of the capacitance:

$$C_g^l(z,t) \approx \underbrace{\bar{C}_g \cdot l}_{C_0} + \underbrace{\bar{C}_g' \int_0^l \phi(x)dx}_{\frac{dC_g}{dz}} \cdot z_d(t) \tag{5.9}$$

where the term $z_d(t)$ represents the harmonic series as shown in Section 3.4. Here, it is written in Euler form for better readability:

$$z_d(t) = z_{d1} \cdot e^{i(\omega t+\psi_1)} + z_{d2} \cdot e^{i(2\omega t+\psi_2)} + ... \tag{5.10}$$

Higher order terms are truncated, because they are too small for a detectable effect. The second order is often truncated as well, but is shown, as it is relevant for potential nonlinear effects. The phases $\psi_1$ and $\psi_2$ of the oscillation are omitted in this derivation and added later again during the fitting procedure of the model.

Expressing the conductance modulation current $i_g$ (Eqn. 5.6) with the modulated charges in the CNT (Eqn. 5.8) yields:

$$i_g = \frac{dG^{DC}}{dV_g^{DC}} \frac{V_{SD}(t)}{C_0} \cdot (C_0 V_{GD}^{AC} \cdot e^{i\omega t} + V_g^{DC} \bar{C}_g' \int_0^l \phi dx \cdot z_d(t)) \tag{5.11}$$

Here it can be seen, that the read-out signal $V_{SD}(t)$ mixes with the AC part of the actuation signal $V_{GD}^{AC} \cdot e^{i\omega t}$ and also with the movement $z_d(t)$ of the oscillating CNT. Depending on the $\omega$ read-out technique ($1\omega$ or $2\omega$), different contributions are present at $\Delta\omega$. The derivation of the components at $\Delta\omega$ depending on the read-out signal are shown in the Appendix (B.1). For the $1\omega$ read-out (Eqn. 5.2), the $\Delta\omega$ component can be separated into a background current $i_{back}$, which is independent of motion, and a motional resonance current $i_{motional}$:

$$i_g^{\omega} = \underbrace{\frac{dG^{DC}}{2 \cdot dV_{GD}^{DC}} \cdot V_{SD}^{AC} \cdot V_{GD}^{AC}}_{i_{\text{back}}} + \underbrace{\frac{dG^{DC}}{dV_{GD}^{DC}} \frac{1}{2C_0} \cdot V_{SD}^{AC} V_{GD}^{DC} \bar{C}_g' \int_0^l \phi dx \cdot z_{d1}}_{i_{motional}} \tag{5.12}$$

It is important to note here, that the two currents $i_{back}$ and $i_{motional}$ have two different phases associated to them. The background current is present independently of the motional current and depends on the transfer characteristic. A verification of the expression is given in Section 5.5.1.

For the $2\omega$ read-out (Eqn. 5.3), the component at $\Delta\omega$ is obtained in the same way (see Appendix B.1). Since the oscillation amplitude in the second harmonic $z_{d2}$ is orders of magnitudes smaller than the first harmonic, this component is often set to zero.

$$i_g^{2\omega} = \frac{dG^{DC}}{dV_{GD}^{DC}} \frac{1}{2C_0} \cdot V_{SD}^{AC} V_{GD}^{DC} \bar{C}_g' \int_0^l \phi dx \cdot z_{d2} \tag{5.13}$$

### 5.3.2 Piezoresitive Current

The piezoresistive current contribution arises from the strain induced resistance change of the CNT. The sensitivity of a material to strain is called gauge factor $G_f$ and defined as:

$$G_f = \frac{\Delta R/R_0}{\Delta l/l} = \frac{\Delta R/R_0}{\epsilon} \tag{5.14}$$

where $R_0$ is the initial resistance that is present at an initial strain $\epsilon_0$ and $\Delta R$ the resistance change and $\epsilon$ is the strain. The change in resistance $\Delta R$ from an initial resistance $R_0$ at strain $\epsilon_0$ is due to a change in length $\Delta\epsilon$:

$$\Delta R = G_f \cdot R_0 \cdot \Delta\epsilon \tag{5.15}$$

The piezoresistive current can then be expressed as:

$$i_p = \frac{V_{SD}(t)}{R_0 + \Delta R} = \frac{V_{SD}(t)}{R_0(1 + G_f \cdot \Delta\epsilon)} \tag{5.16}$$

For small strain changes ($G_f \Delta\epsilon << 1$) this expression can be approximated with a first order Taylor expansion:

$$\frac{1}{x+1} \approx 1 - x \text{ for } x << 1 \tag{5.17}$$

The piezoresistive current can then approximated as:

$$i_p = \frac{V_{SD}}{R_0} (1 - G_f \cdot \Delta\varepsilon) \tag{5.18}$$

The time-dependent strain changes $\Delta\varepsilon$ are only due to the oscillation of the CNT. Pre-strains and previously axially applied strains are not measured at the lock-in amplifier directly, however, they affect the initial resistance $R_0$. The induced tension due to oscillation was derived in Section 3.3.1 and combined with $T = EA\epsilon$ it gives:

$$\Delta\varepsilon = \frac{z^2(t)}{2L} \int_0^L \left(\frac{\partial \phi}{\partial x}\right)^2 dx \tag{5.19}$$

$$\text{with } z(t) = z_s + z_{d1} e^{i(1\omega t + \psi_1)} + z_{d2} e^{i(2\omega t + \psi_2)} \tag{5.20}$$

Merging the expression of the piezoresistive current $i_p$ (Eqn. 5.18) and the change in strain due to oscillation of the CNT (Eqn. 5.19) gives:

$$i_p = \frac{V_{SD}(t)}{R_0} \left(1 - G_f \cdot \frac{z(t)^2}{2L} \int_0^L \left(\frac{\partial \phi}{\partial x}\right)^2 dx\right) \tag{5.21}$$

The current components at $\Delta\omega$ for the two different read out techniques can be separated in the same way as for the conductive current and is shown in Appendix B.2:

$$i_p^{\omega} = -\frac{V_{SD}^{AC} G_f}{R_0 l} \cdot \int_0^L \left(\frac{\partial \phi}{\partial x}\right)^2 dx \cdot \left(\frac{z_{d1} z_s}{2} + \frac{z_{d1} z_{d2}}{4}\right) \tag{5.22}$$

$$i_p^{2\omega} = -\frac{V_{SD}^{AC} G_f}{R_0 l} \cdot \int_0^L \left(\frac{\partial \phi}{\partial x}\right)^2 dx \cdot \left(\frac{z_{d1}^2}{8} + \frac{z_1 z_{d2}}{2}\right) \tag{5.23}$$

At $1\omega$ read out, the current contribution requires either a static gate bias or a strong enough actuation AC signal to trigger a relevant second harmonic.

The gauge factor of CNT arises from the energy bandgap dependent resistances. If strain is applied on the CNT, the bandgaps change and therefore the resistance. The strain dependent bandgap energy changes are highest for SGS CNTs that could lead to gauge factors of over 1000 at large strain levels [147]. Metallic CNTs do not exhibit such changes and will not show this current modulation. Further reading is provided in [148].

## 5.4 Lorentzian Peak Shapes and Fitting Procedure

The lock-in amplifier reads out the the modulated signal at the TIA from the drain electrode and the externally mixed signal as reference. Between both signals there is a phase difference $\phi_{lock-in}$ that is measured by the lock-in amplifier. This phase value does not carry any physical meaning, as the phase of the reference signal is distorted by the pre-amplifier.

The current contributions of the modulated current have different phases. The background current $i_{back}$ has an associated phase of $\Delta\phi$, defined by the electronics, while the motional current due to the oscillation of the CNT has a different phase $\psi$ originating from mechanical motions of oscillators.

A fit for current measurements at the lock-in therefore need to include these degrees of freedom for a successful fit:

$$i^{\Delta\omega} = \left(i_{back} \cdot e^{i\Delta\phi} + i_{motional} \cdot e^{i\psi}\right) \cdot e^{i\phi_{lock-in}} \tag{5.24}$$

To capture the resonance behaviour of the motional current, the motional current is modelled using a normalized Lorentzian function, which follows the mechanical resonance model and keeps a clear physical interpretation of the fitting parameters such as the quality factor $Q$ and the eigenfrequency [58].

$$H = \frac{\omega_0^2}{Q \cdot \left(\omega_0^2 - \omega^2 - i\frac{\omega \cdot \omega_0}{Q}\right)} \tag{5.25}$$

The motional current can then be fitted with:

$$i_{motional} = |i_{motional}| \cdot H \tag{5.26}$$

The fitting procedure (Eqn. 5.24) allows for the rapid extraction of the eigenfrequency and the quality factor. An example of device CGDS 21 is shown in Figure 5.3. The presented shape of the resonance peak differs significantly and is explained by the more dominant phase of the background current $i_{back}$, when the motional current $i_{motional}$ is low for lower gate biases.

However, the quality factor $Q$ is not reliably fitted with this method. In both Figure 5.3a and 5.3b, it is visible, that the fit is disturbed by the noise. Another approach to fit the quality factor is the ring-down experiment, which

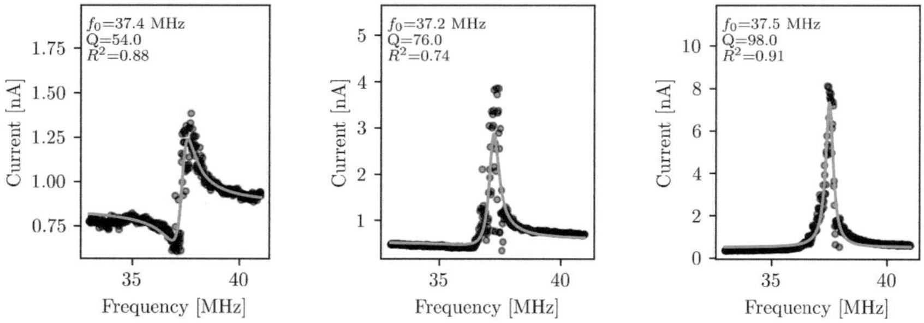


**(a)** $V_{GD}^{DC} = 3\,\text{V}$. **(b)** $V_{GD}^{DC} = 4.5\,\text{V}$. **(c)** $V_{GD}^{DC} = 6\,\text{V}$.

**Figure 5.3:** Three different frequency sweeps on device CGDS 21 at different gate biases ($V_{GD}^{AC} = 10\,\text{mV}, V_{SD}^{AC} = 40\,\text{mV}$). The shape of the Lorentzian peak changes, because the amplitude of the motional current increases and dominates the total phase stronger.

was not performed, as the frequency down-mixing technique is not a time-resolved detection method.

The fitting was automised with python using a custom `pytorch` model for gradient optimization which extracted phases better than other approaches, for example `scipy.optimize`.

## 5.5 Electrical Measurement Results

Parts of this section were published in [79, 80]. In those publications, the CNTs were modelled as a solid cylinders. Here in this work, the CNT is remodelled with a tube geometry, leading to different estimation in strain levels in the CNT.

After identifying a potentially working device (with resistance preferably $<1\,\text{M}\Omega$ and small bandgap semiconducting, as described in Section 5.1), the first step is to identify the eigenfrequency as well as finding first good working parameters for gate and source AC. Most eigenfrequencies with expected pre-strains for the used suspended length lie between 15 MHz and 100 MHz.

To achieve a a full measurement set, both the gate bias and the actuation frequency were swept. For each gate bias in an alternating symmetric sequence ($V_{GD}^{DC} = 0\,\text{V}, 1\,\text{V}, -1\,\text{V}, 2\,\text{V}, -2\,\text{V}...$) a frequency sweep was performed. The frequency sweep is typically executed in the upward direction, although the setup allows for the sweep direction to be reversed or performed bidirectionally.

For each step of the frequency sweep, a new gate actuation frequency is applied, after which the system pauses for a duration equivalent to 7 times the time constant at the lock-in amplifier, ensuring that the system reaches a steady state. Following this stabilization period, measurement data are acquired, and the actuation frequency is subsequently updated for the next step in the sweep.

Figure 5.4 shows a full overview over such measurement results on device CGDS 21 which are analysed step by step. The results of a second device (SJDS 123) of the same architecture is shown in the Appendix D.2. These devices did not have thermal actuators and were fabricated by Dr. Seoho Jung and Cristina Gentili (suspended length $l = 2\,\mu\text{m}$, $g_0 = 2\,\mu\text{m}$).

Both $1\omega$ and $2\omega$ read-out techniques were successful to measure the eigenfrequencies of the CNT. With increasing gate bias $V_{GD}^{DC}$, the eigenfrequency of this devices increases from 37.4 MHz up to 41 MHz following the same trend predicted by the gate bias tuning discussed in Section 3.6. As expected from calculations of the current contributions in Section 5.3.1 and 5.3.2, the current amplitude of the $1\omega$ read out technique is significantly higher than for the $2\omega$ read-out technique.

In a next step, the Lorentzian fitting is performed for $1\omega$ and $2\omega$ for each gate bias. The data and fits are shown for gate biases of 8 V and −8 V in Figure 5.6. Here it can be seen that the $2\omega$ current is much lower then the $1\omega$ current.

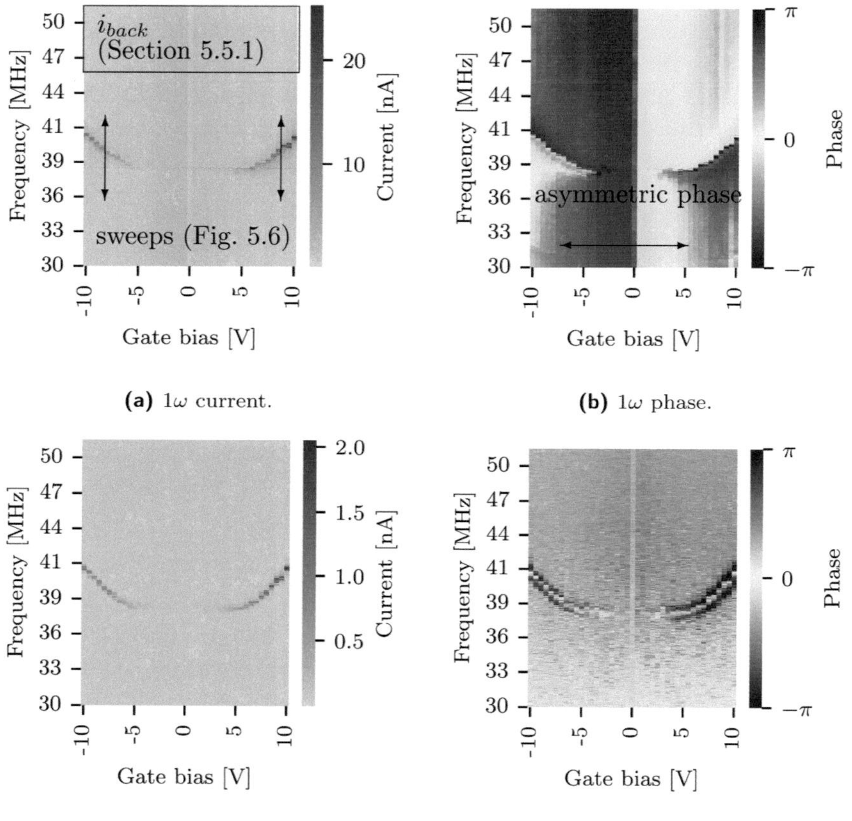


**(a)** $1\omega$ current.

**(b)** $1\omega$ phase.

**(c)** $2\omega$ current.

**(d)** $2\omega$ Phase.

**Figure 5.4:** Full result overview of device CGDS 21. showing current amplitude and phase for different gate biases and both $\omega$ read out techniques with $V_{GD}^{AC} = 10\,\text{mV}$, $V_{SD}^{AC} = 40\,\text{mV}$. The eigenfrequency was tuned from 37.4 MHz up to 41 MHz.

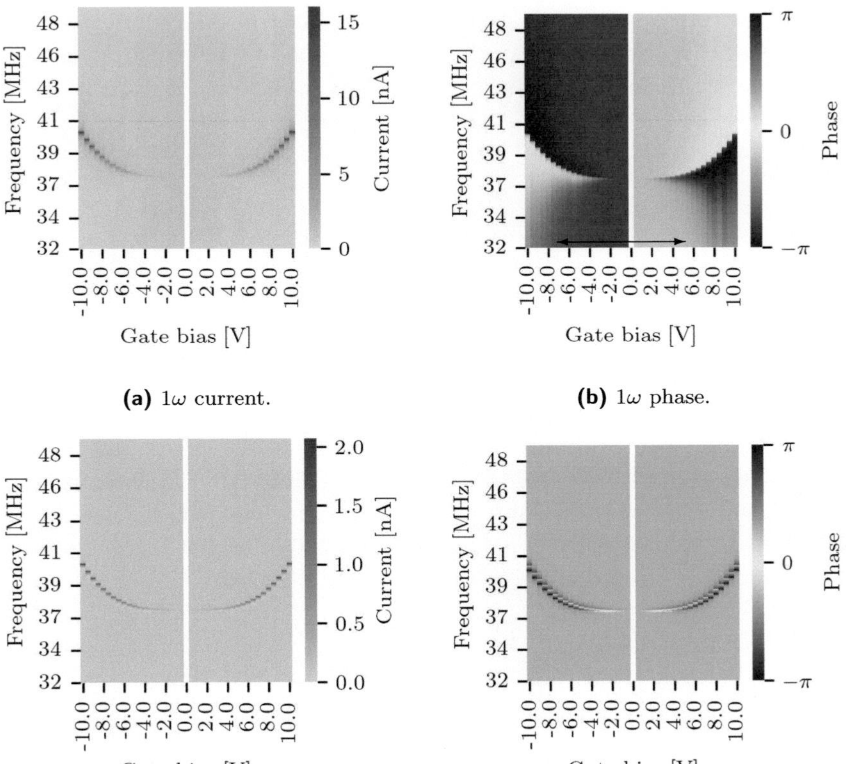


**(a)** $1\omega$ current.

**(b)** $1\omega$ phase.

**(c)** $2\omega$ current.

**(d)** $2\omega$ Phase.

**Figure 5.5:** Modelled current and phase responses of device CGDS 21.

### 5.5.1 Background Current Analysis

The Lorentzian fit does not include any information about the current conduction physics presented in Section 5.3.1 and 5.3.2. Here, the validity of the expression for the back ground current in the $1\omega$ read-out technique (as derived in Eqn. 5.12) is analysed.

Five frequency sweeps have been conducted at different gate AC values (ranging between 7.5 mV and 20 mV) and the background currents have been extracted over the range of 25 MHz above the eigenfrequencies as indicated in Figure 5.4a. The averaged currents are shown in Figure 5.7a and show increasing currents for increasing gate AC, as expected from Equation 5.12.

Equation 5.12 can be rearranged to:

$$\frac{2 \cdot i_{back}}{V_{SD}^{AC} \cdot V_{GD}^{AC}} = \frac{dG^{DC}}{dV_{GD}^{DC}} \tag{5.27}$$

This relation is used to verify the measurement technique, as $i_{back}$ is measured at the lock-in amplifier in a dynamic current measurement, whereas $\frac{dG^{DC}}{dV_{GD}^{DC}}$ is measured in a different measurement before and after the sweep with static power sources. The two different measurements are shown in Figure 5.7a, where the background current $i_{back}$ is scaled according to Equation 5.27 and the absolut value of the transconduction $\frac{dG^{DC}}{dV_{GD}^{DC}}$ is plotted. The measurements match well in shape and amplitude.

### 5.5.2 Current Modelling of CNTs with Harmonic Balancing

Instead of relying only on the fitting approach from the Lorentzian with phase, both the resonance behaviour and the current behaviour are modelled from bottom up using the equations presented in Chapter 3 and the current transconduction explained in this chapter. The extracted eigenfrequencies from the Lorentzian fits are used to perform the fit of the mechanical model according to the overview in Figure 3.9. Parameters radius $r$, mass-loading $\alpha_{ML}$, and pre-strain $\epsilon_0$ are fitted until the base frequency at 0 V gate bias as well as all frequency tuned by the gate bias match. Then the solution of the HB model contains static deflection $z_s$, dynamic amplitudes $z_{d1}$ and $z_{d2}$, and the phase of the mechanical motions $\psi_1$, $\psi_2$. Then, the current model is applied with the modelled deflections. The result is shown in Figure 5.5.

The same modelling steps that were shown in this chapter for device

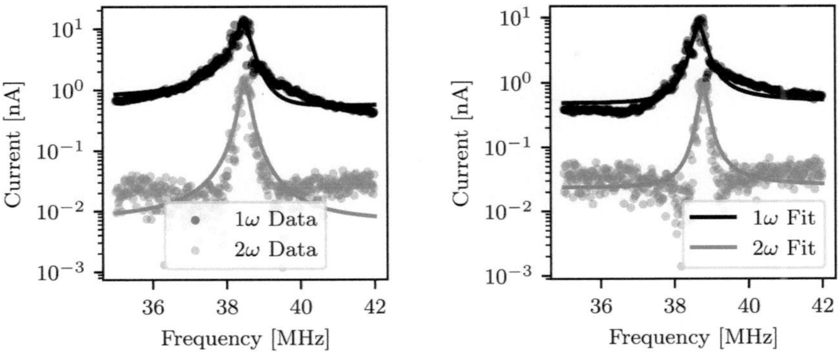


**(a)** $V_{GD}^{DC} = 8\,\text{V}$.

**(b)** $V_{GD}^{DC} = -8\,\text{V}$.

**Figure 5.6:** Overlay of $1\omega$ and $2\omega$ measurements of device CGDS 21 with $V_{GD}^{AC} = 10\,\text{mV}$, $V_{SD}^{AC} = 40\,\text{mV}$. The background current of the $2\omega$ read-out technique is two orders of magnitude smaller than the background current of the $1\omega$ read-out. For both read-outs, the eigenfrequency is detected is the same ($f_0 = 38.5\,\text{MHz}$ at $V_{GD}^{DC} = 8\,\text{V}$, $f_0 = 38.7\,\text{MHz}$ at $V_{GD}^{DC} = -8\,\text{V}$).

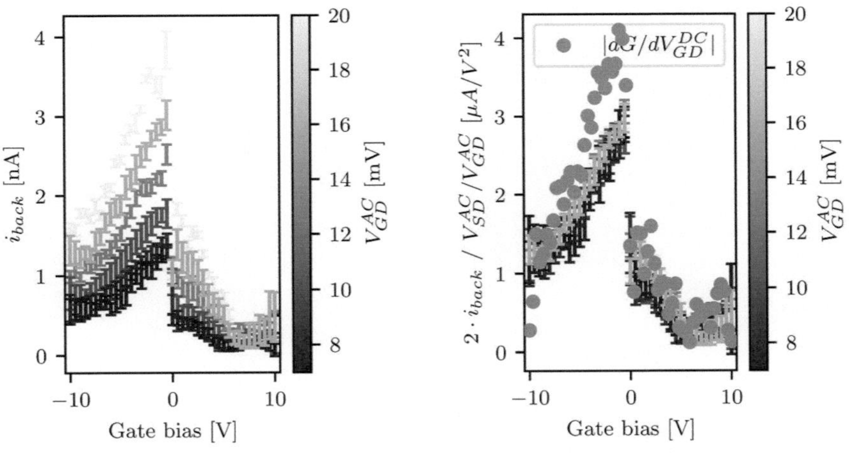


**(a)** Averaged background current data as shown in Figure 5.4a for different gate ac $V_{GD}^{AC}$. They scale linearly as expected from Equation 5.12.

**(b)** Normalized background current (measured by the lock-in) by $V_{SD}^{AC}$, $V_{GD}^{AC}$, multiplied with 2 and the absolute value of the transconductance $dG/dV$ (Fig. 5.1b).

**Figure 5.7:** Comparison of the back-ground current as measured by the lock-in amplifier and the static transfercharacteristics as measured by the Keithleys on CGDS 21.

**Table 5.1:** Values used for modelling of devices CGDS 21 and SJDS 123. Radius $r$, mass-loading factor $\alpha_{ML}$ and pre-strain $\epsilon_0$ are fitted during the HB mechanical model. Once the eigenfrequency matched, the Gauge factor $G_f$ was adjusted to fit the current amplitude.

| | Device | CGDS 21 | SJDS 123 |
|---|---|---|---|
| Fixed | $l$ | 2.1 µm | 2 µm |
| | $g_0$ | 2 µm | 2 µm |
| | $E$ | 1 TPa | 1 TPa |
| | $h$ | 340 pm | 340 pm |
| HB fit | $r$ | 1.1 nm | 0.8 nm |
| | $\alpha_{ML}$ | 6 | 2.3 |
| | $\epsilon_0$ | 0.03 % | 0.04 % |
| Current fit | $G_f$ | 1100 | 620 |

CGDS 21, were also performed on SJDS 123 and are shown in the Appendix D.2.

The gauge factor is used as a fitting parameter. For device CGDS 21, the gauge factor was fitted to be 1100, and for SJDS 123 620 (summarized in Tab. 5.1). Both values have to be treated with caution. While the current model allowed for the inclusion of the conductance modulation $dG/dV$ to be gate bias dependent, the model used one constant quality factor and one constant gauge factor for all gate biases. Additionally, device CGDS 21 had 6 CNTs transferred, making it impossible to attribute exact contribution to any tube.

Another potential explanation is addressed in Section 6.3. There are multiple combinations of parameters ($\epsilon_0$, $r$, $\alpha_{ML}$) that can fit the mechanical behaviour of the single data sets presented in this Section. While the gate bias induced tension on the CNT remains the same, the resulting deflections can vary drastically, which in turn influences the current amplitudes.

However, despite missing information, the model successfully captured the asymmetry and the different shapes of the Lorentzian peak as well as replicating complex phases.

The achievements of the modelling procedure are summarized below:

- The current amplitude of the $1\omega$ data becomes only detectable at around $\pm 2.5$ V gate bias, because the motional current amplitude is too small. Similar observations are done for the second device measured (SJDS 123, see Appendix D.2). SJDS 123 shows a clear asymmetry between negative and positive gate biases.
- This asymmetry is even more visible in the phase of the $1\omega$ current for

both devices. The ability to replicate phase data with the model is a novelty, which has been made possible through the application of HB.

- For $2\omega$ measurement, the model is also capable of replicating both the phase and the amplitude of the current. The current amplitude of the $2\omega$ model is smaller than the measured current.

# 6 Results

This chapter presents an experimental investigation into the strain response of CNT resonators. Using the previously characterized MEMS actuators, axial strain was applied to suspended CNTs to tune their eigenfrequencies. Both pure axial strain and a combination of gate bias-induced strain and axial strain were explored.

In the case of pure axial strain (with no applied gate voltage), the fundamental eigenfrequency of the resonator could be modelled using the simple beam-string equation (see Eqn. 2.1). Fitting this model to experimental data enabled the extraction of key parameters, such as the cross-sectional radius $r$, mass-loading factor $\alpha_{ML}$, and pre-strain $\epsilon_0$.

Whereas the previous chapter focused on modelling electrical currents and frequency tuning via gate bias on devices without axial strain capability, this chapter expands the analysis by introducing axial straining via the MEMS platforms introduced in Chapter 4. The focus here lies on eigenfrequency tuning and mechanical modelling of CNT resonators under strain.

The first part of the chapter discusses experiments involving pure axial straining. These serve as a simplified scenario, allowing for model validation without the added complexity of gate bias effects. These experiments are only feasible thanks to the newly designed platform presented in this thesis. This approach offers direct insight into the applied strain levels and validates the employed continuum model. Measurements were conducted on three devices differing in CNT length and actuator configuration.

In the second part, the validity of the extracted parameters ($r$, $\alpha_{ML}$, and $\epsilon_0$) is tested against additional experiments that combine gate bias-induced strain with axial strain. The resulting strain response is modelled using the Duffing equation, incorporating the previously fitted mechanical parameters. In particular the radius, as extracted from pure axial strain, needs readjustment to fit the influence of the gate bias tuning.

Additionally, slippage events were observed during the experiments, providing insight into the mechanical adhesion strength of the CNT clamps.

A discussion is presented, contextualizing the obtained parameter values and give suggestions regarding the underlying physics.

## 6.1 Pure Axial Strain Tuning

This section will analyse the suspended CNT resonance behaviour due to axial strain and without any gate bias induced strain in an experimental setup. For this analysis, results from three different thermal actuator devices are available, covering all actuator designs (A6, A9, A13) and two different suspended CNT lengths (2.9 µm and 4.2 µm).

In comparison to the setup presented in Section 5.2, the setup for the axially strainable chips, required two extra power sources to drive the thermal actuators. The adapted setup is shown in Figure 6.1.

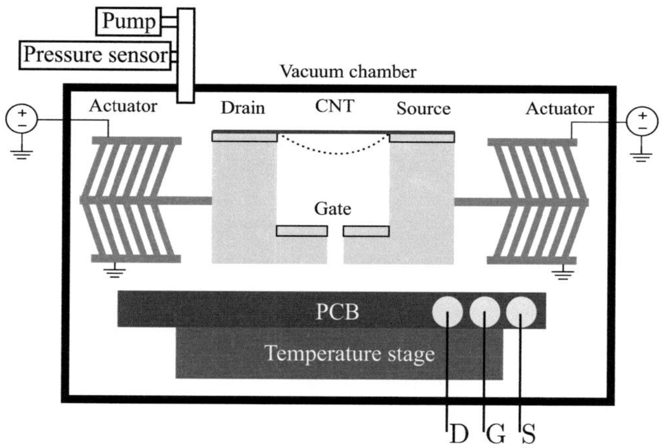


**Figure 6.1:** Adapted measurement setup for straining experiments. Both thermal actuators require a power source to control the applied strain.

The measurement of pure axial strain without deflection due to static gate bias $V_{GD}^{DC} = 0\,\text{V}$ has inherent challenges. At 0 V gate bias, the motional current of the conductance modulation current (Eqn. 5.12) becomes zero. The contribution of piezoresistive current (Eqn. 5.22) consists mainly on the product $z_{d1} \cdot z_s$, where for the case of pure axial strain $z_s$ is zero. The second contribution is by the product of $z_{d1} \cdot z_{d2}$, where $z_{d2}$ is usually so small, that it gets neglected often.

For a first measurement performed on device MV 01 (A13, $a = 1.035\,\text{nm}\,\text{mA}^{-2}$, $l = 4.2\,\mu\text{m}$), multiple frequency sweeps were performed at $V_{GD}^{DC} = 0\,\text{V}$, $V_{GD}^{AC} = 10\,\text{mV}$, $V_{SD}^{DC} = 0.4\,\text{V}$, $V_{SD}^{AC} = 35\,\text{mV}$ with $1\omega$

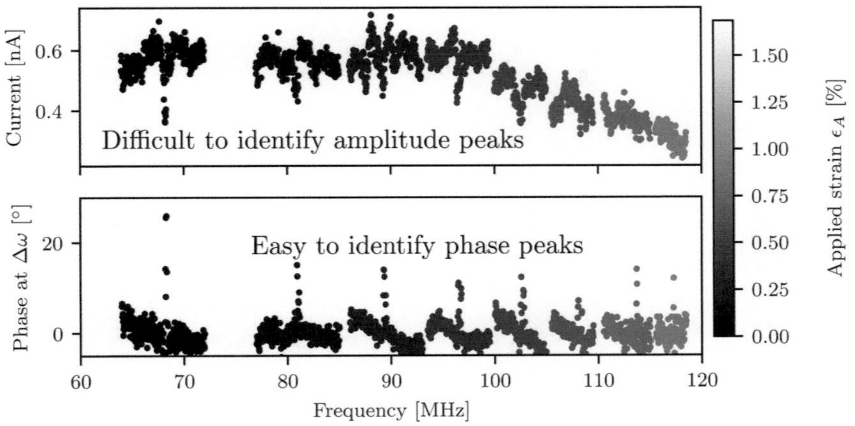


**Figure 6.2:** Eight data sets of device MV 01 truncated around the eigenfrequency at different applied strain levels. Current and phases measured at $V_{GD}^{DC} = 0\,\mathrm{V}$, $V_{GD}^{AC} = 10\,\mathrm{mV}$, $V_{SD}^{DC} = 400\,\mathrm{mV}$, and $V_{SD}^{AC} = 35\,\mathrm{mV}$. For higher strain levels, the resonance peak in the current amplitude gets smaller until it indistinguishable from noise. In the current phase however, the resonance frequency remains identifiable.

read-out technique. The actuator currents $I_{act}$ were swept from 0 mA up to 10 mA. In Figure 6.2, measurement results up to current 6 mA are shown. The current amplitude is dominated by noise and the resonance peak is only visible for some sweeps, as expected, because only piezoresistive current is contributing. The conductance motional current is zero for zero gate bias $V_{GD}^{DC}$. The resonance peaks however, are clearly visible in the phases of the current. To provide a readable figure, all phases were adjusted to that the background phase of the off-resonance current is zero. The eigenfrequencies were extracted at the maximum phase, allowing the creation of data pairs of actuator currents and resonance frequencies ($I_{act}$, $f_0$) forming the basis for further analysis.

The first measured eigenfrequency is 68.3 MHz for 0 mA applied actuator current. It was tuned up to 117.3 MHz for 6 mA, which is a tuning range of 49 MHz or 70 % relative tuning range. At actuator currents of 6 mA, the eigenfrequency drastically dropped down to 93 MHz and then further increased up to 114.8 MHz with increasing actuator current up to 8.3 mA (Fig. 6.3).

This event is attributed to abrupt slippage. The CNT tension got so high, that the adhesively clamped contacts could not withstand it and slipped. For currents higher than $I_{act} = 8.3\,\mathrm{mA}$, the CNT signal was not detectable any longer. A second slippage event is unlikely, because the transfer

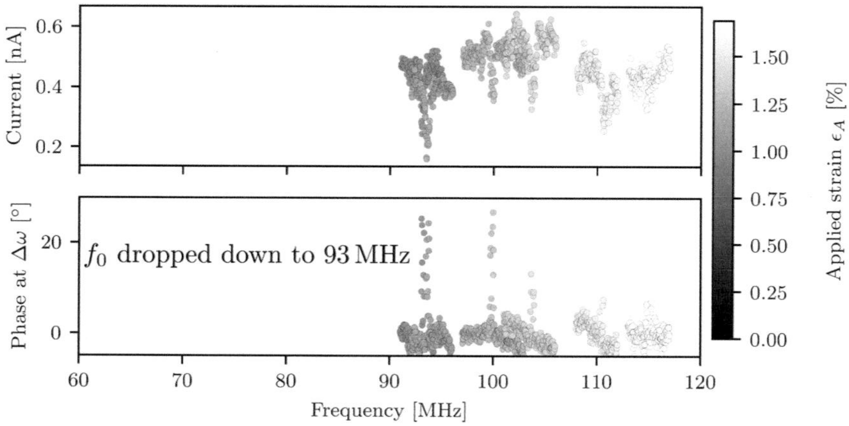


**Figure 6.3:** The eigenfrequency of device MV 01 abruptly decreased down to 93 MHz due to slippage and increased again due to further applied strain.

characteristics (see Section 6.2) continued the expected strain trend. After the measurement series, the CNT showed visible slack in SEM (shown in Appendix D.4).

To understand and estimate the pre-tension and applied tensions by the actuators, the beam-string eigenfrequency behaviour (Eqn. 2.1) is fitted onto the data ($I_{act}$, $f_0$). For this fit, the amount of fitting parameters is reduced to a minimum: the suspended length $l_0 = 4.2\,\mu\text{m}$ was measured in SEM and the gate distance was determined in WLI $g_0 = 490\,\text{nm}$. The displacement coefficient $a = (1.035 \pm 0.020)\,\text{nm}\,\text{mA}^{-2}$ was evaluated using SEM extraction (Tab. 4.5). The Young's modulus is fixed at $E = 1\,\text{TPa}$ and the wall thickness is $h = 340\,\text{pm}$.

The remaining unknowns are pre-strain $\epsilon_0$, mass-loading factor $\alpha_{ML}$, and radius $r$ and will serve as fitting parameters. The cross-sectional area $A_c$, the moment of inertia $I_y$, and the linear mass density $\mu$ are computed from the tube model presented in Section 3.5. The fitting was performed without bound restrictions on the parameters, ensuring that the solution space was not artificially limited. The procedure of this fitting process is summarized in Figure 6.4.

After the slippage event, the radius $r$, length $l$, and mass-loading factor $\alpha_{ML}$ of the CNT are assumed to remain unchanged, and only the pre-strain is expected to reduce. With this assumption, the fitting process takes the fitting quality in form of the $R^2$-value of both fits into account and the fit is chosen by maximizing the product $R^2_{\text{before slip}} \cdot R^2_{\text{after slip}}$. The result of

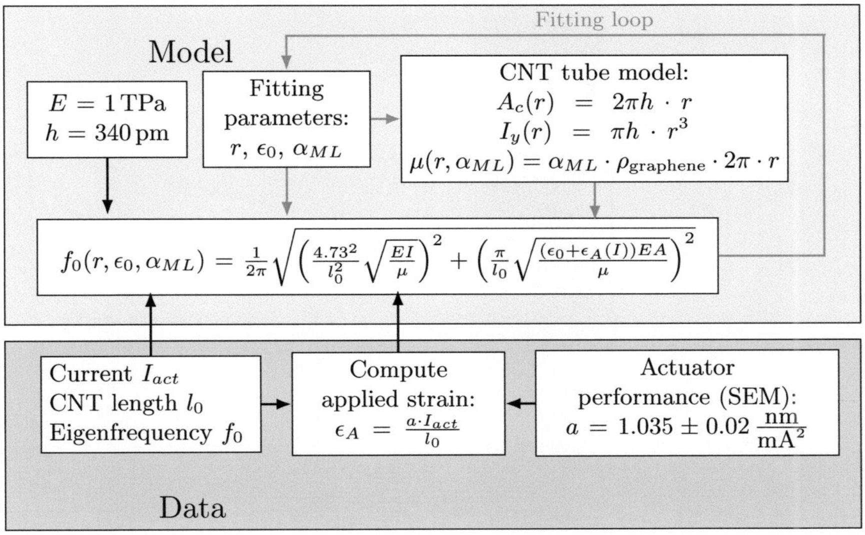


**Figure 6.4:** Fitting procedure to find the radius, pre-strain and line density of the CNT for pure uniaxial strain tuning. Experimental data are tuples of $I_{act}$ and $f_0$, the characterized actuator displacement and the measured CNT length.

this fit is shown in Figure 6.5 based on the model parameter values given in Table 6.1.

The results are shown as the eigenfrequency versus the applied strain of the thermal actuator. As a reminder at this point, the strain of the CNT consists of a potential pre-strain and the applied strain. If the CNT is slacked, a certain amount of strain needs to be applied, before the eigenfrequency equation (Eqn. 2.1) yields real results. The internal CNT strain $\epsilon_i$ is expressed as the superposition of initial strain $\epsilon_0$ and applied strain $\epsilon_A(I_{act})$.

$$\epsilon_i(I_{act}) = \epsilon_0 + \epsilon_A(I_{act}) \tag{6.1}$$

The applied strain $\epsilon_A(I_{act})$ of the actuator can not be negative. In Equation 2.1, the pre-strain present in the CNT can shift the curve towards left, if $\epsilon_0 > 0$. This can be seen in Figure 6.5 before the slippage. The pre-strain can be interpreted as the x-axis shift from the origin and is indicated with a gray dashed line.

The strain compositions are broken down with arrows in Figure 6.5. The black arrow represents the applied strain $\epsilon_A$, the dashed arrow the pre-strain $\epsilon_0$ if positive or the slack to be removed if negative. The solid coloured arrow

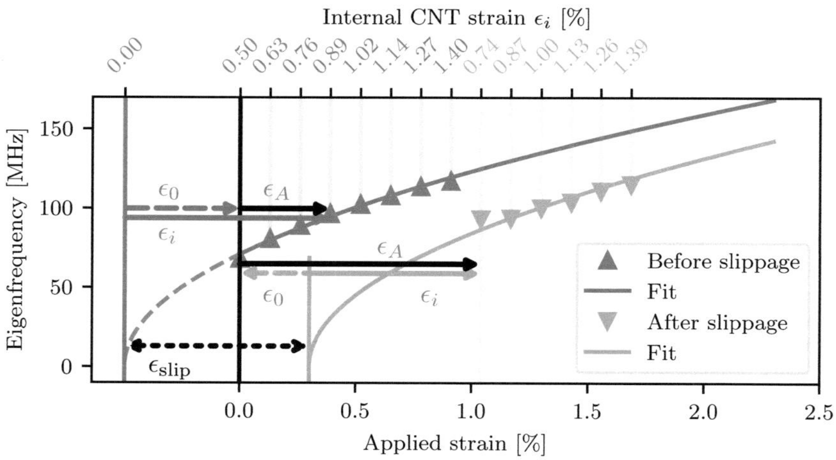


**Figure 6.5:** The extracted eigenfrequencies of MV 01 were fitted with Equation 2.1 for $r$ and $\alpha_{ML}$, but with different pre-strains $\epsilon_0$. The difference of the individual pre-strains is the lost strain due to slippage $\epsilon_{slip}$. The arrows visualize Equation 6.1.

**Table 6.1:** Fitted parameters from pure axial strain experiments. The reported errors arise form the covariance matrix returned by the `scipy.optimize` fitting routine and using an uncertainty of the eigenfrequencies of $\sigma_{f0} = 1\,\text{kHz}$. The displacement coefficient $a$ of device MV 39 was not measured, but estimated.

| | **Parameter** | **MV 01** | **MV 12** | **MV 39** |
|---|---|---|---|---|
| Actuator | $a$ [nm mA$^{-2}$] | $1.035 \pm 0.03$ | $0.984 \pm 0.052$ | $2.77 \pm 0.015$* |
| | Design | A13 | A9 | A6 |
| CNT | $l$ [µm] | 4.2 | 2.9 | 2.9 |
| | $E$ [TPa] | 1 | 1 | 1 |
| | $h$ [pm] | 340 | 340 | 340 |
| Fitted | $\alpha_{ML}$ | $6.23 \pm 0$ | $5.61 \pm 0$ | $6.08 \pm 0$ |
| | $\epsilon_0$ [%] | $0.50 \pm 0.01$ | $-1.14 \pm 0.01$ | $-0.66 \pm 0.01$ |
| | $r$ [nm] | $0.68 \pm 20$ | $0.65 \pm 3.11$ | $0.74 \pm 11.9$ |
| | $R^2$ | 0.989 | 0.999 | 0.999 |

shows the total strain $\epsilon_i$ present in the CNT, which determines the resonance behaviour.

If $\epsilon_0 < 0$ in Equation 2.1 the curve is shifted towards right. While in physics, negative strain means compression, in this equation it indicates slack at zero applied strain. Effectively, the negative pre strain values is what the actuator has to apply for the CNT to have total internal strain $\epsilon_i \geq 0$. This is shown for device MV 01 in Figure 6.5 after slippage.

The fit shows, that the CNT first had a pre-strain of $\epsilon_0 = (0.50 \pm 0.01)\,\%$, and then slipped after a maximum internal strain of 1.4 %, losing 0.8 % internal strain, which brought the CNT into a slacked regime, if the applied strain is removed. This was confirmed via SEM and is shown in Appendix D.4.

This device (MV 01) has been presented in [81], where the reported values, which were derived from the simulated applied strain $\epsilon_A$, differ from the ones presented here. This difference will be discussed in Section 6.3.

Two other devices (MV 12 and MV 39) were also analysed with this process. Other than for device MV 01, the eigenfrequencies were not detectable at 0 V gate bias, neither in current amplitude nor in the current phase. They only became detectable at $V_{GD}^{DC} = 2\,\mathrm{V}$ for device MV 12 and $V_{GD}^{DC} = 5\,\mathrm{V}$ for device MV 39. To estimate the eigenfequencies at $V_{GD}^{DC} = 0\,\mathrm{V}$, the Duffing equation was fitted using HB in the same way as for the devices in Chapter 5 using the well identifiable eigenfrequencies at larger gate biases (schematic from Fig. 3.9). This procedure was only performed to get a good estimation of the eigenfrequencies at 0 V gate bias.

The eigenfrequencies were then used for the same fitting procedure as in Figure 6.4. The results are presented in Figure 6.6. The fitting results shown in Table 6.1 stem from the fitting process in Figure 6.4.

Other than for device MV 01, the CNTs on both device MV 12 (A9) and device MV 39 (A6) exhibited initial slack and required currents of $I_{act} = 6\,\mathrm{mA}$ and $I_{act} = 3\,\mathrm{mA}$ respectively, to remove slack. From the fit in Figure 6.6 applied strain from the actuators were 1.14 % strain for device MV 12 and 0.68 % strain for device MV 39. Then device MV 12 was tuned from 43.3 MHz up to 110.1 MHz ($\Delta f = 66.8\,\mathrm{MHz}$, 154 % relative range) and device MV 39 from 63.5 MHz up to 120.9 MHz ($\Delta f = 57.4\,\mathrm{MHz}$, 85.9 % relative range). The CNT on device MV 12 slipped at 0.52 % internal strain and lost 0.06 % strain.

The uncertainty in $\alpha_{ML}$ and the pre-strain $\epsilon_0$ in Table 6.1 are relatively small, which gives confidence in the extracted values. The third fitting parameter, namely the radius $r$, exhibits very large uncertainty, order of magnitudes larger, than the extracted value itself. From a practical point of view this data set is unsuitable to fit the radius, as it apparently does not contain information about it. On the other hand, this can be well

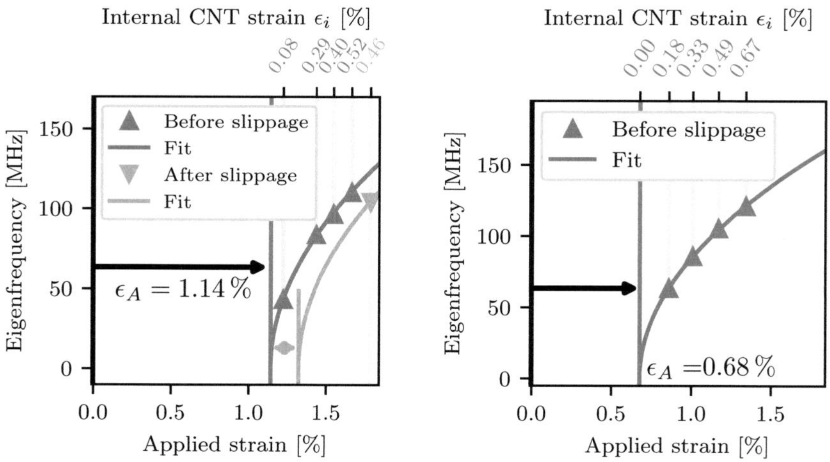


**(a)** Extracted eigenfrequencies at 0 V gate bias of device MV 12 with an A9 actuator design and $l = 2.9\,\mu m$.

**(b)** Extracted eigenfrequencies at 0 V gate bias of device MV 39 with an A6 actuator design and $l = 2.9\,\mu m$.

**Figure 6.6:** Fitting of mass-loading factor and pre-strain of devices MV 12 and MV 39.

understood, if we notice, that all the measurement points are in the string regime of the CNT, because of the high tensile strain levels. The radius influence vanishes, because in the string term (Eqn. 2.1) both the cross-sectional area (Eqn. 3.73) and the linear mass density (Eqn. 3.72) depend linearly on the radius, which therefore cancels out. To extract the radius from the uniaxial strain experiments reliably, more data points in a low strain regime would be required.

In Section 6.3, we show extended data sets of both device MV 12 and MV 39, that also contain a gate bias influence. In that section, we re-asses if those data sets can be utilized to get the CNT radius.

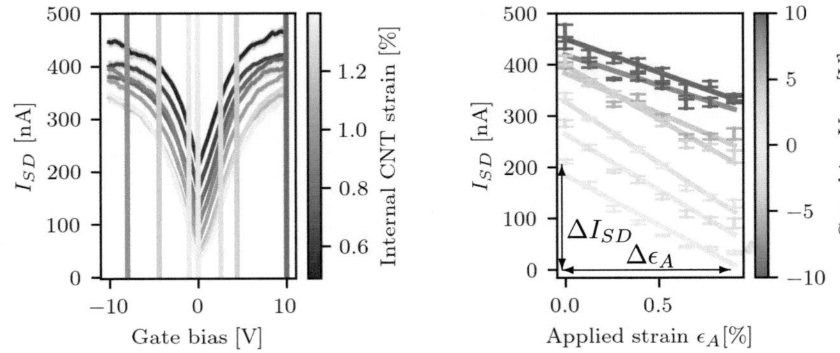


**(a)** Strain dependent transfer characteristics of device MV 01. The source-drain current decreases for increasing applied strain indicating a positive gauge factor. The vertical lines represent the gate bias values used for gauge factor extraction by slope.

**(b)** Gauge factor extraction by the slope of the source drain currents shown for different gate biases.

**Figure 6.7:** Static current response of MV 01.

## 6.2 Influence of the Strain on the Transfer Curve

The strain does not just influence the mechanical response of the CNT, but also its electrical conductivity through the effect of piezoresistivity. The change in conductivity due to strain was measured in the transfer characteristics of the CNT. Figure 6.7a shows the transfer-curve of device MV 01 before slippage at $V_{SD}^{DC} = 300\,\text{mV}$ for different internal strains. The strain values were computed from the fitted pre-strain (Fig. 6.5) and the applied strains from the thermal actuator. The source current $I_{SD}$ decreases with increasing applied strain, indicating a positive gauge factor.

The static gauge factor was extracted by analysing the slope of changing source currents with respect to strain and initial current $I_0$, which is equivalent to the gauge factor definition (Eqn. 5.14). This process is shown in Figure 6.7b. For the range of applied strain, the slope is approximated to be linear. This approximation works well for larger gate bias amplitudes. For lower gate biases, the linear approximation is worse.

$$G_f = \frac{-\Delta I_{SD}}{I_{SD}(\epsilon_A = 0\,\%) \cdot \Delta\epsilon_A} \tag{6.2}$$

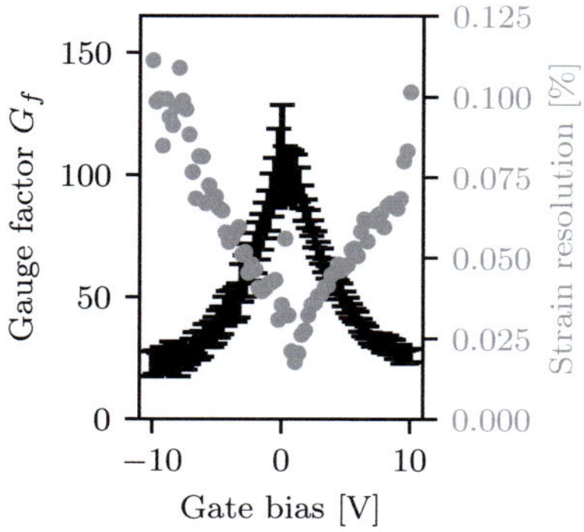


**Figure 6.8:** Extracted gauge factors of device MV 01 before the slippage of the CNT. The gauge factors reach over 100 for low gate bias and reduce down to 22 for larger gate bias. The strain resolution is best for lower gate biases going down to 0.02 % strain.

The gauge factors are shown in Figure 6.8. The gauge factors are over 100 (largest extracted is 108.4) for low gate biases and reduce for higher gate biases down to 22. The noise level of the measured currents $\sigma_{ISD}$ is used to determine the strain resolution of the static strain gauge. The strain sensitivity of the source-drain current $S_{ISD}$ is the slope of the linear fits shown in Figure 6.7b. The strain resolution is then $\sigma_{ISD}/S_{ISD}$ and is also shown in Figure 6.8. The best strain resolution is 0.02 % at low gate bias, where the gauge factors are largest.

After the slippage the system behaves in the same way as shown in Figure 6.8. The trend of the reducing current continues, even for actuator currents higher than $I_{act} > 8.3\,\text{mA}$. At $I_{act} = 8.3\,\text{mA}$, the resonance current was not detectable neither in amplitude nor in phase, suggesting a second slippage event. However, a continuing trend in the transfer characteristics (Fig. 6.9), suggests continuous straining without a second slippage.

## 6.2.1 Influence of the Mass-loading Factor

Device MV 01 has been presented in [81], where the reported values of $G_f$ and $\epsilon$ differ from the ones presented here. At the time of the study [81], the thermal actuators have not been characterized via SEM as described in Section 4.3. More importantly, the model utilized in the study did not include a mass-loading factor. This omission has led to an underestimation of the strain values, which in turn has affected the estimation of the gauge factors.

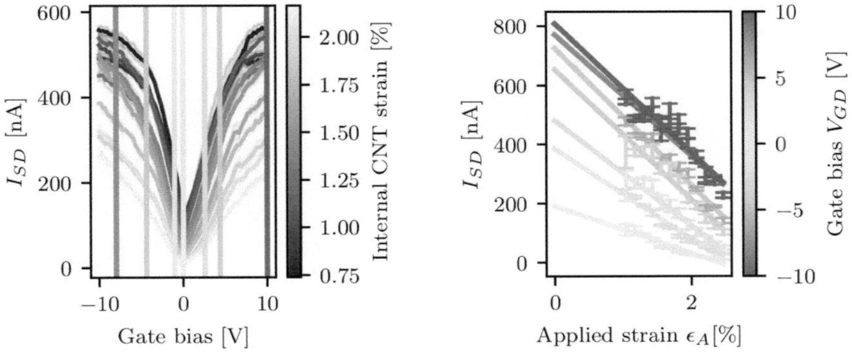


**(a)** Strain dependent transfer curves. **(b)** Slope extraction of the gauge factor.

**Figure 6.9:** Transfer curves of device MV 01 after the slippage event.

To understand the influence of the mass-loading factor, we rewrite Equation 2.1 into the form below.

$$f_{0,n} = \frac{1}{2\pi}\sqrt{\frac{E \cdot h}{\alpha_{ML}\rho_g}}\sqrt{\underbrace{\left(\frac{\lambda_n^2}{l^2}\cdot\sqrt{\frac{r^2}{2}}\right)^2}_{\text{Beam}} + \underbrace{\left(\frac{n\cdot\pi}{l}\cdot\sqrt{\epsilon_A(I_{act})+\epsilon_0}\right)^2}_{\text{String}}} \tag{6.3}$$

neglecting the beam influence under higher tension and rearranging:

$$f_{0,n} = \frac{n}{2l}\sqrt{\frac{E\cdot h}{\rho_g}}\sqrt{\frac{\epsilon_A(I_{act})+\epsilon_0}{\alpha_{ML}}} \tag{6.4}$$

As can be seen in this equation, the internal strain can be divided by the mass-loading factor $\alpha_{ML}$.

In the absence of SEM measurements of the actuator displacement, estimating the applied strain was done using a COMSOL simulation. This simulation confirmed the low strain values in that study. Upon closer investigation, we realized that this accidental confirmation of this model was due to unchecked value of electrical resistivity.

As discussed in Section 4.4, the resistivity used in this thesis is based on semiconductor physics, yielding a room temperature value of $14.4\,\mathrm{m\Omega\cdot cm}$, which agrees with the wafer manufacturer's specification ($< 20\,\mathrm{m\Omega\cdot cm}$). The IEEE MEMS model used a resistivity which was too low for the specification of the manufacturer ($4\,\mathrm{m\Omega \cdot cm}$) (Fig. 6.10a). Unfortunately, no independent resistivity measurements have been performed on the wafer.

Another key difference concerns the geometry of the simulation. The IEEE MEMS model was a 3D simulation, whereas the thesis model employed a 2D approach. As a result, the treatment of heat flux into the handle layer differed. In the 3D model, the handle layer was included and its bottom surface was set to ambient temperature as a boundary condition. However, it did not model the BOX layer and therefore lacked thermal interface resistance effects. The 2D thesis model instead used out-of-plane heat flux as a fitting parameter.

Most importantly, the silicon resistivity was lower at lower temperatures leading to less generated heat and less displacement. Other changes that were made include the semiconducting behaviour of silicon (with the doping as fitting parameter), which only becomes relevant for higher temperatures.

When all these effects are combined, the displacement vs. actuator current characteristic, reveals a discrepancy factor of around 10 (Fig. 6.10b). This discrepancy is then carried over into the internal strain and gauge factor estimations.

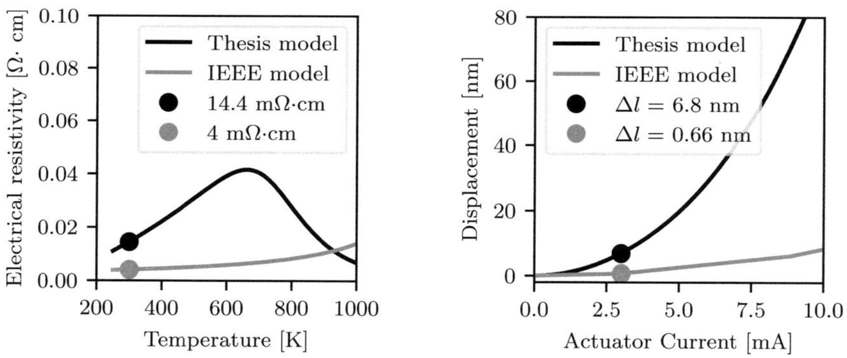


**(a)** Electrical resistivity comparison.

**(b)** Displacement comparison.

**Figure 6.10:** Key difference comparison between the models used.

**Table 6.2:** Differences in the values from the IEEE MEMS 2024 analysis in comparison with this work.

| Parameter | MEMS 2024 [81] | This thesis |
|---|---|---|
| Mass-loading factor $\alpha_{ML}$ | 1 | 6.23 |
| Pre-strain $\epsilon_0$ | 0.44 m$\epsilon$ | 0.5 % |
| $\epsilon_{\text{slip}}$ | 0.44 m$\epsilon$ | 0.8 % |
| Highest $G_f$ | 1060 | 108.4 |
| Static resolution | 61 $\mu\epsilon$ | 0.02 % |

## 6.3 Combination of Axial Strain and Gate Bias Tuning

In this Section, measurements combining axial strain and gate bias strain are presented. The combined straining effects are modelled with the Duffing equation presented in Section 3.3.3, which is solved with the HB method presented in Section 3.4.

As model inputs, the values for the mass-loading factor $\alpha_{ML}$ and pre-strain $\epsilon_0$ are taken from Table 6.1. Those values were obtained from the pure axial strain data sets. As discussed in Section 6.1, the value of radius of the CNT can not be obtained from those data sets. Therefore, in this Section, we investigate the influence of the radius on the gate bias tuning in combination with axial strain experiments.

For both devices (MV 12 and MV 39), four data sets were available at different actuator currents $I_{act}$ containing both actuation frequency sweeps and gate bias sweeps. The four data sets for device MV 12 before slippage are shown in Figure 6.11 and for device MV 39 in Figure 6.12 (both truncated to a frequency range of interest).

The device MV 12 has much better visible eigenfrequencies and even shows the second mode in its first measurement. Device MV 39 is much more noisy. Both devices however show very similar behaviour. At lower applied strain the eigenfrequency exhibits a large tunability with gate bias tuning the eigenfrequencies from 44 MHz up to 91 MHz and ≈63.5 MHz up to ≈92 MHz. For larger applied strains the eigenfrequencies increase, but the influence of the gate bias tuning reduces to be less than 10 MHz. This behaviour is expected and has been shown in Figure 3.8.

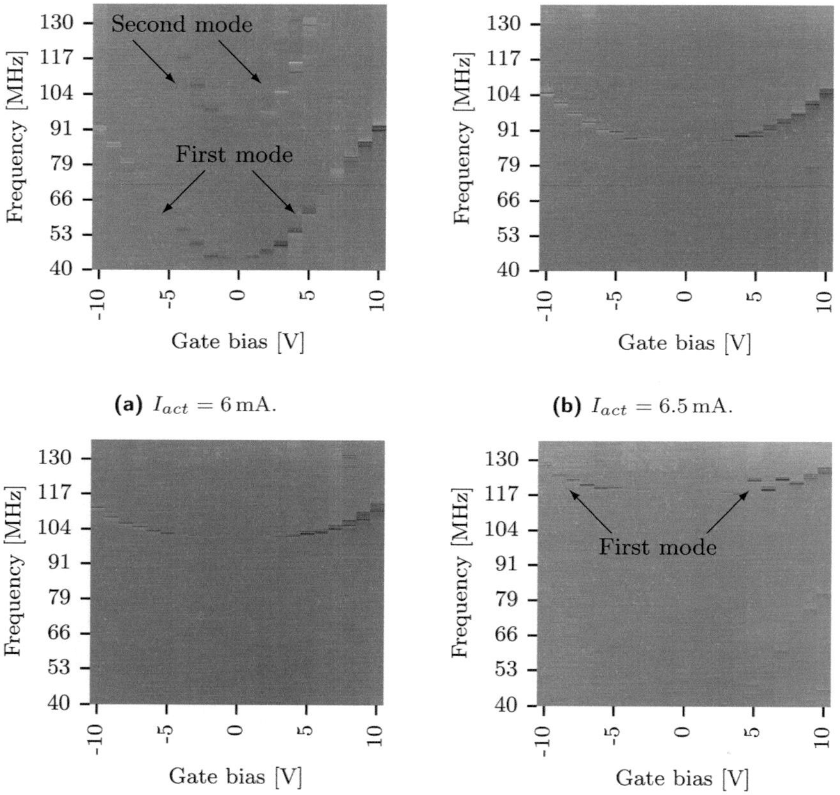


**(a)** $I_{act} = 6\,\mathrm{mA}$.

**(b)** $I_{act} = 6.5\,\mathrm{mA}$.

**(c)** $I_{act} = 6.75\,\mathrm{mA}$.

**(d)** $I_{act} = 7\,\mathrm{mA}$.

**Figure 6.11:** Four datasets at different applied actuator currents of device MV 12 (A9). For $I_{act} = 6\,\mathrm{mA}$ the measurement also captures the second mode of the CNT. For increasing actuator currents, the eigenfrequencies increase. The tuning range from the applied gate bias decreases with higher axially applied strain.

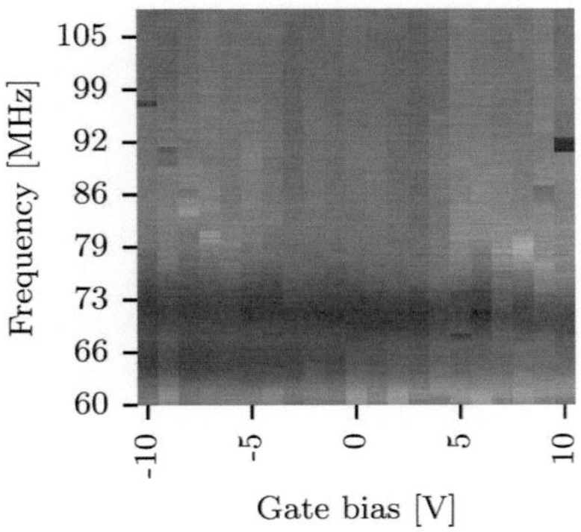


**(a)** $I_{act} = 3.0\,\mathrm{mA}$.

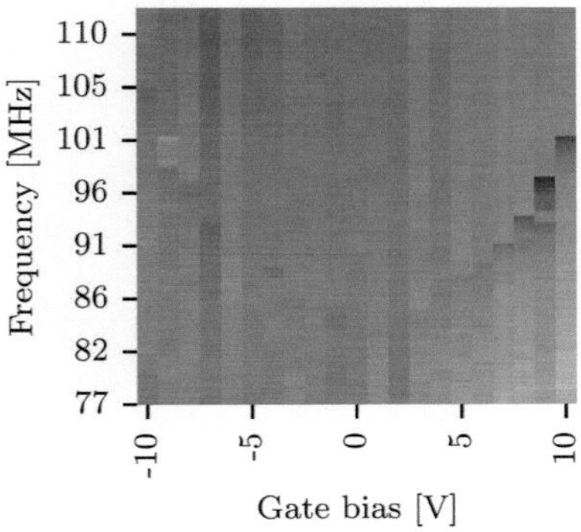


**(b)** $I_{act} = 3.25\,\mathrm{mA}$.

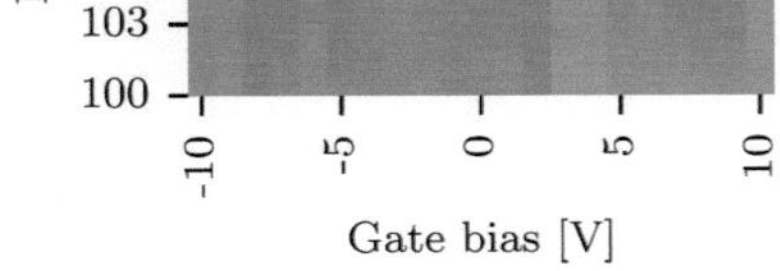


**(c)** $I_{act} = 3.5\,\mathrm{mA}$.

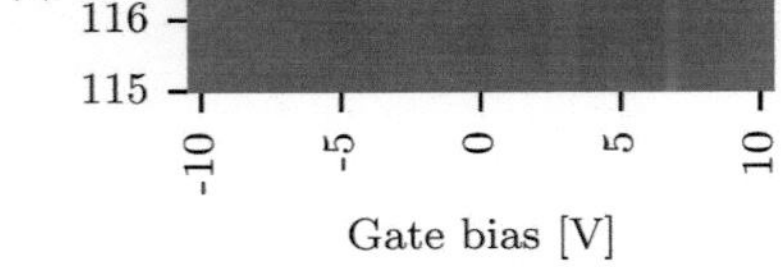


**(d)** $I_{act} = 3.75\,\mathrm{mA}$.

**Figure 6.12:** Four datasets at different applied actuator currents $I_{act}$ of device MV 39 (A6). The eigenfrequencies are much more difficult to detect and are more noisy, especially for actuator currents $I_{act} > 3.5\,\mathrm{mA}$. This device shows very similar behaviour to device MV 12.

Eigenfrequencies were extracted using the Lorentzian fit from Section 5.4. The HB model is then applied for different CNT radii. The results for MV 12 are shown in Figure 6.13, similar results for MV 39 are shown in Appendix D.5.

For reasonable radius values of around 1 nm (dashed line in Fig. 6.13, the model fails to match the extracted eigenfrequencies. Figure 6.13 shows, that in order to match the gate bias influence well, the fit requires a radius of $r = 0.08\,\text{nm}$ (continuous line in Fig. 6.13). This unrealistically small value challenges the validity of the pure mass-loading hypothesis.

To address this issue, we perform HB analysis based on different combination of parameters. As shown in Figure 6.14, both the mass-loading factor and the product $E \cdot h$ are reduced by the same factor and then a radius was fitted. One set is done with $\alpha_{ML} = 1$ in which case $E \cdot h$ has to be reduced by a factor of 6.51 for a reasonable radius of 0.75 nm. It is also possible to reduce $E \cdot h$ by a smaller factor 5, which leads to an $\alpha_{ML} = 1.12$ for a radius of 0.7 nm.

With respect to the pure axial strain data set, according to the pre-factor of Equation 6.3, the factor of 6.51 can be clearly redistributed between $E \cdot h$ and $\alpha_{ML}$ without changing the eigenfrequency at zero gate bias. However, the data set presented in this Section, that contains gate bias tuning, shows that the contribution of uniaxial strain and gate bias strain are different. To support this statement, we rewrite Equation 2.1 to explicitly include the gate bias tension $T_g$.

$$f_{0,n} = \frac{1}{2\pi}\sqrt{\left(\frac{\lambda_n^2}{l^2} \cdot \sqrt{\frac{EI_y}{\mu}}\right)^2 + \left(\frac{n \cdot \pi}{l} \cdot \sqrt{\frac{(\epsilon_0 + \epsilon_A(I_{act}) \cdot EA_c + T_g}{\mu}}\right)^2} \tag{6.5}$$

Note that, this equation is not explicitly solved for eigenfrequencies, these are obtained by HB, but this simple model can help us to understand the different influences of the two straining methods. While the MEMS is applying strain directly, the gate bias tuning is applying an electrostatic force. To a first order, the tension in the CNT $T_g$ has to balance this electrostatic force in statics. It is therefore independent on the stiffness. The tension introduced by the MEMS is stiffness dependent ($T = EA\epsilon$). The resonance frequency in the string regime is then finally dictated by the ratio of tension to mass.

This reveals that there is an important distinction between the two straining modes. While in the pure axial strain experiments ($T_g = 0$), the measurements where independent of the radius, now the radius influences the tension to mass ratio of the gate bias tuning through the mass-term.

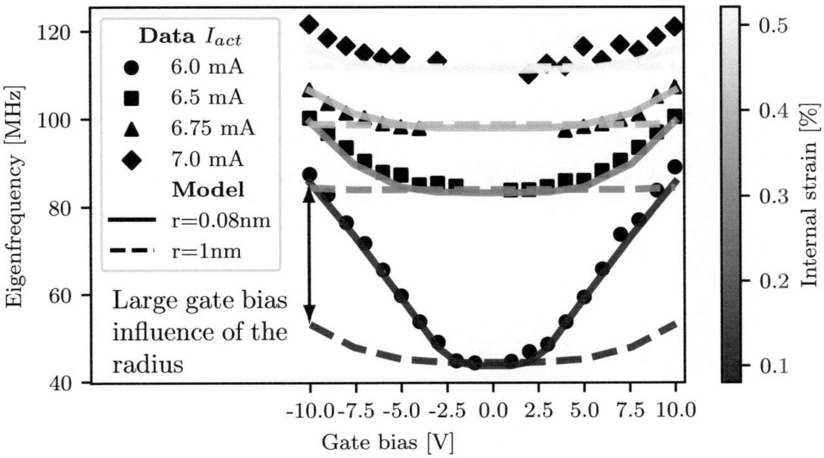


**Figure 6.13:** Gate bias influence displayed for two different radii with the the previously fitted pre-strain $\epsilon_0 = -1.14\,\%$ and $\alpha_{ML} = 5.61$. The radius $r = 0.08\,\text{nm}$ that replicates the gate bias tuning behaviour well is unrealistically small, while a realistic radius ($r = 1\,\text{nm}$) underestimates the gate bias tuning.

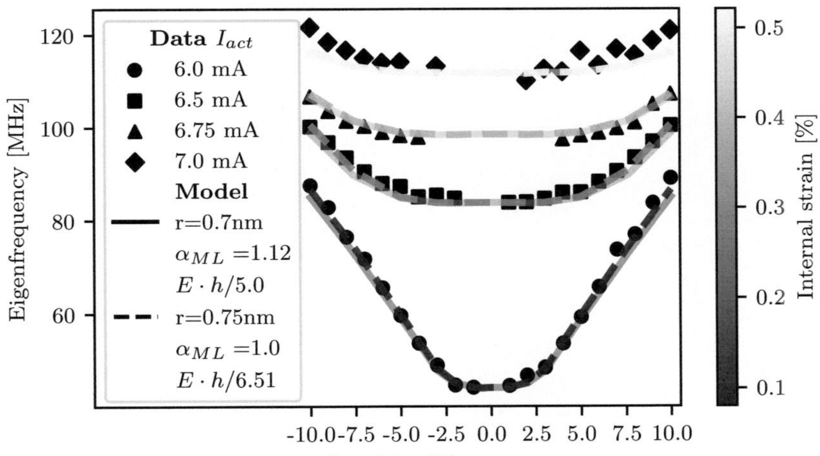


**Figure 6.14:** Other HB simulation results for device MV 01. The model can match the data points well, when $E \cdot h$ and $\alpha_{ML}$ are reduced by the same factor. The fitting radii are more realistic than the previous fit.

While it is still possible to find combinations of stiffness and mass-loading that match the gate bias tuning data sets, there is an asymmetry between the importance of stiffness and mass-loading. In the pure uniaxial data sets, this was ambiguous. Here, there is a slight preference for the stiffness over mass-loading, because, dividing $E \cdot h$ by a factor of 6.51 and $\alpha_{ML} = 1$ yields good fits of both data sets, whereas assigning 6.51 to $\alpha_{ML}$ and leaving $E \cdot h$ unchanged, only fits the uniaxial strain data set and is therefore inconsistent.

As seen in Table. 2.2, this product $E \cdot h$ usually lies between $300\,\mathrm{N\,m^{-1}}$ and $400\,\mathrm{N\,m^{-1}}$. The only realistic values, which are $\alpha_{ML} > 1$ and $r > 0.7\,\mathrm{nm}$ require a $E \cdot h$ in the range $50\,\mathrm{N\,m^{-1}}$.

While we can not conclude further based on the existing data on mechanisms that could reduce the longitudinal stiffness, this could generally be attributed to either the wall thickness being smaller, the Young's modulus being smaller or an imperfect transfer of strain from the electrodes to the CNT.

Independent determination of the radius would have been desirable, but was not measured, because of the fragility of the devices.

# 7 Conclusions and Outlook

## 7.1 Summary and Conclusions

This thesis investigated the tuning of carbon nanotube resonators by applying strain through two primary methods - gate bias and axial strain - and explored a combined approach which is capable of applying both. The study aimed to validate new modelling techniques, to fabricate micro-electro-mechanical systems (MEMS) capable of precisely applying axial strain, and to demonstrate the feasibility of significantly modifying CNT resonant frequencies.

### Gate-Bias Tuning and Harmonic Balancing Modelling

Initial experiments focused on rigid device architectures where suspended CNTs could be strained electrostatically by applying a gate bias. This setup allowed an evaluation of a modelling framework based on the nonlinear Duffing equation, solved by harmonic balancing. By analysing the device's mixing currents, the model captured not only the asymmetric current amplitudes with respect to the gate bias but also, for the first time, the phase behaviour of these mixing currents. The harmonic balancing approach successfully explains the observed resonance shapes and subtle differences between positive and negative gate biases, which was a key finding. Although this verification was limited to electrostatic strain, the ability of the model to accurately reproduce both amplitude and phase characteristics shows promise for broader applications, including more complex tuning mechanisms.

### Axial Strain with a Novel MEMS Platform

To achieve pure axial strain, a custom-designed MEMS actuator was employed. Joule heating in chevron-style expansion beams displaced the clamping ends of the CNT, enabling controlled tensile strain. Three distinct actuator designs (with 6, 9, or 13 beams) were fabricated, each characterized via scanning electron microscopy (SEM). The axial actuation platform allowed application of pure strain to the CNT without the deflection induced by electrostatic forces, yielding a measured strain of up to 1.4%, and resonant frequencies could be tuned by over 150%, with some devices reaching 124 MHz. The tuning range was ultimately limited by the

CNT-electrode contact strength, highlighting the importance of robust mechanical and electrical interfaces; enhanced clamping strategies (e.g., top-side clamping or improved adhesion layers) could further expand the feasible strain range.

### Combined Gate-Bias and Axial Tuning

The thesis also investigated a combined approach that superimposed electrostatic (gate) strain onto an axially strained CNT. While pure uniaxial strain experiments could not give information about radius, but mainly pre-strain, the combined straining state allowed for deeper analysis. When both tuning mechanisms were applied together, the results indicated that a reduced longitudinal stiffness of the CNT was required to match the experimental data. This finding underscores the necessity of precise material parametrization in CNT resonator models.

### Overall Significance

By integrating advanced modeling with a versatile MEMS actuation platform, this work demonstrated high-frequency, large-range CNT resonator tuning based on well-characterized axial and gate-bias-induced strain. Moreover, the continuum model used for the CNT's structural properties offers a refined framework for future investigations of CNT electromechanical phenomena. The combination of new MEMS structures, detailed SEM-based strain measurements, and rigorous modeling approaches provides a robust toolkit for CNT resonator research. Although strains of up to 1.4% were achieved, further improvements in contact strength will be necessary to approach the ultimate performance limits. In addition, fully exploiting the promise of CNT resonators requires stronger evidence of long-term device stability and reproducibility, which calls for larger sample sizes and extended measurement campaigns.

## 7.2 Outlook

Building upon these results, there are several directions for future research that can extend the applicability and enhance the performance of tunable CNT resonators:

1. **Stronger and More Reliable CNT-Electrode Interfaces**
   Improving the mechanical and electrical contact remains critical for realizing higher strain levels without CNT slippage. Potential approaches include top-clamping techniques, advanced adhesion layers, or local surface treatments at the contact regions. Strengthening the interface would not only expand the tuning range but also improve device robustness under repeated strain cycles.

2. **Shorter CNT Lengths for Higher Frequencies**
   While this work achieved resonant frequencies in the 100+ MHz range with suspended lengths around 2.9 - 4.2 µm, employing shorter CNTs is a straightforward way to push resonance frequencies even higher. Coupled with robust mechanical interfaces, such devices could open paths toward GHz-range resonators suitable for radio-frequency applications.

3. **Versatile Platforms for Other 2D Materials**
   The MEMS platform developed here can be adapted for other nanomaterials, such as graphene or molybdenum disulfide ($MoS_2$). Investigating how axial and electrostatic strains affect vibrational modes of these 2D crystals may lead to a new generation of tunable resonators with different mechanical and electrical properties.

4. **Automation and Pre-Characterization**
   The startup endeavors mentioned in this thesis show promising steps toward industrial-level application by automating CNT transfer and using Raman spectroscopy for pre-characterization. Combining such high-throughput assembly methods with the versatile strain platforms demonstrated here could eventually yield large datasets needed to optimize device fabrication and performance.

5. **Quality Factor Enhancement and Operation at Ambient Conditions**
   Increasing strain can lead to damping dilution, potentially boosting the quality factor enough to operate under ambient pressure. This capability is crucial for real-world applications, such as sensors or filters, where vacuum conditions are not always practical. Further research is required to fine-tune the strain-induced quality factor enhancements and to assess long-term device stability outside controlled laboratory environments.

6. **Potential for Industrial Adoption**
   Although the first tunable CNT resonator demonstration dates back about two decades, industrial-scale deployment remains elusive. Continued improvements in fabrication reliability, contact strength, and device uniformity-together with robust, automated assembly-could accelerate the integration of CNT resonators in commercial applications, from tunable radio-frequency filters to high-sensitivity sensors.

Through addressing these challenges, future work can build upon the platform and findings presented in this thesis. The potential for large-frequency-range tuning, combined with the fundamental insights into CNT structural properties, holds promise for enabling a new class of highly tunable and miniaturized mechanical resonators. With continued progress, CNT-based technologies may yet see broad adoption, fulfilling the vision that sparked the early enthusiasm for these remarkable nanomaterials.

# A Mechanical Model: Supplementary Material

## A.1 Galerkin's Method for the Motional PDE

The motional PDE (Eqn. 3.36) was derived to:

$$\mu\phi_n(x)\frac{\partial^2 z(t)}{\partial t^2} + EI_y z(t)\frac{\partial^4\phi_n(x)}{\partial x^4} - Tz(t)\frac{\partial^2\phi_n(x)}{\partial x^2} + b\phi_n(x)\frac{\partial z(t)}{\partial t} = f(x,t) \tag{A.1}$$

The solution of this equation was given in Equation 3.6 as

$$z(x,t) = \sum_{n=1}^{\infty}\phi_n(x)\cdot z_{0,n}\cos(\omega_n t) \tag{A.2}$$

The idea of Galerkin's method is to project the original PDE onto a single mode shape at a time, by multiplying Equation 3.36 with one specific mode shape $\phi_n(x)$. This projection of a continuous system onto a lower-dimensional system converts the PDE into an ODE for each mode. This works because the mode shapes are orthogonal [53, 149]:

$$\int_0^l \phi_m(x)\phi_n(x)dx = 0, \quad \text{for} \quad m \neq n \tag{A.3}$$

The orthogonality ensures that the integration decouples each mode. The multiplication with $\phi_n(x)$ yields:

$$\mu\phi_n^2(x)\frac{\partial^2 z(t)}{\partial t^2} + EI_y z(t)\phi_n(x)\frac{\partial^4\phi_n(x)}{\partial x^4} - Tz(t)\phi_n(x)\frac{\partial^2\phi_n(x)}{\partial x^2} + b\phi_n^2(x)\frac{\partial z(t)}{\partial t} = f(x,t)\phi_n(x) \tag{A.4}$$

And the integration over the length $l$ of the beam:

$$\int_0^l \left( \mu\phi_n^2(x)\frac{\partial^2 z(t)}{\partial t^2} + EI_y z(t)\phi_n(x)\frac{\partial^4 \phi_n(x)}{\partial x^4} \right.$$
$$\left. -Tz(t)\phi_n(x)\frac{\partial^2 \phi_n(x)}{\partial x^2} + b\phi_n^2(x)\frac{\partial z(t)}{\partial t} \right) dx = \int_0^l f(x,t)\phi_n(x)dx \quad \text{(A.5)}$$

The individual terms of this equation are treated separately. The inertia and the damping term give:

$$\int_0^l \mu\phi_n^2(x)\frac{\partial^2 z(t)}{\partial t^2} dx = \mu\frac{\partial^2 z(t)}{\partial t^2} \cdot \int_0^l \phi_n^2(x)dx \quad \text{(A.6)}$$
$$\int_0^l b\phi_n^2(x)\frac{\partial z(t)}{\partial t} dx = b\frac{\partial z(t)}{\partial t} \cdot \int_0^l \phi_n^2(x)dx \quad \text{(A.7)}$$

The bending stiffness term:

$$\int_0^l EI_y z(t)\phi_n(x)\frac{\partial^4 \phi_n(x)}{\partial x^4} dx = EI_y z(t) \int_0^l \phi_n(x)\frac{\partial^4 \phi_n(x)}{\partial x^4} dx \quad \text{(A.8)}$$

is further simplified with integration by parts twice. The rule of integration by parts is:

$$\int_0^l uv'dx = [u(x)v(x)]_0^l - \int_0^l u'vdx \quad \text{(A.9)}$$

$$\int_0^l \underbrace{\phi_n(x)}_{u}\underbrace{\frac{\partial^4 \phi_n(x)}{\partial x^4}}_{v'} dx = \left[ \underbrace{\phi_n(x)}_{u} \cdot \underbrace{\frac{\partial^3 \phi_n(x)}{\partial x^3}}_{v} \right]_0^l - \int_0^l \underbrace{\frac{\partial \phi_n(x)}{\partial x}}_{u'}\underbrace{\frac{\partial^3 \phi_n(x)}{\partial x^3}}_{v} dx \quad \text{(A.10)}$$

The boundary conditions (Equations 3.24 and 3.25) eliminate the the term $u \cdot v$, because $\phi_n(x) = 0$ for $x = 0$ and $x = l$, yielding:

$$\int_0^l \underbrace{\phi_n(x)}_{u}\underbrace{\frac{\partial^4 \phi_n(x)}{\partial x^4}}_{v'} dx = -\int_0^l \underbrace{\frac{\partial \phi_n(x)}{\partial x}}_{u'}\underbrace{\frac{\partial^3 \phi_n(x)}{\partial x^3}}_{v} dx \quad \text{(A.11)}$$

The integration by parts is again applied on the right side of this equation:

$$\int_0^l \underbrace{\frac{\partial \phi_n(x)}{\partial x}}_{u} \underbrace{\frac{\partial^3 \phi_n(x)}{\partial x^3}}_{v'} dx = \left[ \underbrace{\frac{\partial \phi_n(x)}{\partial x}}_{u} \cdot \underbrace{\frac{\partial^2 \phi_n(x)}{\partial x^2}}_{v} \right]_0^l - \int_0^l \underbrace{\frac{\partial^2 \phi_n(x)}{\partial x^2}}_{u'} \underbrace{\frac{\partial^2 \phi_n(x)}{\partial x^2}}_{v} dx \tag{A.12}$$

Again, the boundary conditions (Equations 3.24 and 3.25) eliminate the the term $u \cdot v$, because $\frac{\partial \phi_n(x)}{\partial x} = 0$ for $x = 0$ and $x = l$. This yields the expression:

$$EI_y z(t) \int_0^l \phi_n(x) \frac{\partial^4 \phi_n(x)}{\partial x^4} dx = EI_y z(t) \int_0^l \left( \frac{\partial^2 \phi_n(x)}{\partial x^2} \right)^2 dx \tag{A.13}$$

The axial tension term in Equation A.5 is treated in the same way:

$$\int_0^l -Tz(t) \phi_n(x) \frac{\partial^2 \phi_n(x)}{\partial x^2} dx = -Tz(t) \int_0^l \phi_n(x) \frac{\partial^2 \phi_n(x)}{\partial x^2} dx \tag{A.14}$$

applying the rule of integration by parts here:

$$\int_0^l \underbrace{\phi_n(x)}_{u} \underbrace{\frac{\partial^2 \phi_n(x)}{\partial x^2}}_{v'} dx = \left[ \underbrace{\phi_n(x)}_{u} \underbrace{\frac{\partial \phi_n(x)}{\partial x}}_{v} \right]_0^l - \int_0^l \underbrace{\frac{\partial \phi_n(x)}{\partial x}}_{u'} \underbrace{\frac{\partial \phi_n(x)}{\partial x}}_{v} dx \tag{A.15}$$

giving:

$$\int_0^l -Tz(t) \phi_n(x) \frac{\partial^2 \phi_n(x)}{\partial x^2} dx = Tz(t) \int_0^l \left( \frac{\partial \phi_n(x)}{\partial x} \right)^2 dx \tag{A.16}$$

Inserting the terms A.6, A.7, A.13, and A.16 into Equation A.5 gives the lumped element Equation 3.37 as presented in Section 3.3:

$$\begin{aligned} \mu \int_0^l \phi_n^2(x) dx \frac{\partial^2 z(t)}{\partial t^2} + b \int_0^l \phi_n^2(x) dx \frac{\partial z(t)}{\partial t} + EI_y z(t) \int_0^l \left( \frac{\partial^2 \phi_n(x)}{\partial x^2} \right)^2 dx \\ + Tz(t) \int_0^l \left( \frac{\partial \phi_n(x)}{\partial x} \right)^2 dx = \int_0^l f(x,t) \phi_n(x) dx \end{aligned} \tag{A.17}$$

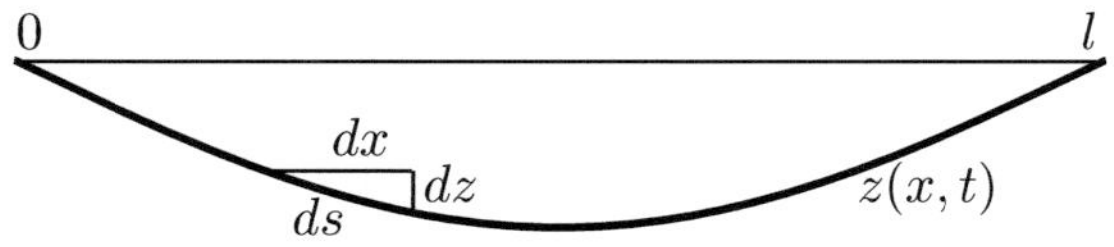


**Figure A.1:** Schematic of the beam elongation due to deflection.

## A.2 Beam Elongation due to Transverse Displacement

Consider a straight, undeflected beam of initial length $l$. When the beam undergoes a transverse deflection $z(x,t)$, its actual (arc) length increases to $l + \Delta l$ due to geometric stretching.

Let the infinitesimal arc length be denoted by $ds$ as sketched in Figure A.1. From the Pythagorean theorem we derive:

$$ds = \sqrt{dx^2 + dz^2} = \sqrt{1 + \left(\frac{dz}{dx}\right)^2}\quad dx \tag{A.18}$$

(A.19)

The total deformed length $l'$ of the beam is then given by:

$$l' = l + \Delta l = \int_0^l \sqrt{1 + \left(\frac{dz}{dx}\right)^2}\, dx \tag{A.20}$$

For small deflections, where $(dz/dx)^2 \ll 1$, a Taylor expansion yields:

$$\sqrt{1 + \left(\frac{dz}{dx}\right)^2} \approx 1 + \frac{1}{2}\left(\frac{dz}{dx}\right)^2 \tag{A.21}$$

The change in length $\Delta l$ compared to the original length is:

$$\Delta l = l' - l = \int_0^l \frac{1}{2}\left(\frac{dz}{dx}\right)^2 dx \tag{A.22}$$

Thus, the axial strain due to elongation becomes:

$$\epsilon = \frac{\Delta l}{l} = \frac{1}{2l}\int_0^l \left(\frac{dz}{dx}\right)^2 dx \tag{A.23}$$

Assuming linear elasticity, the corresponding axial tension $T$ from Hooke's law is:

$$T = \sigma A = EA\epsilon \tag{A.24}$$

Substituting the strain $\epsilon$ (Eqn. (A.23)) into Equation A.24 yields:

$$T = \frac{EA}{2l}\int_0^l \left(\frac{dz}{dx}\right)^2 dx \tag{A.25}$$

which describes the axial tension $T$ due to transverse displacement as induced by the static gate bias, described in Section 3.3.1.

## A.3 Harmonic Balancing Equations

In this section, the equations yielded by the HB method, required for solving the differential Equation 3.62 to obtain the static deflection $z_s$ and the dynamic amplitudes of deflection $z_{dn}$. Let $T = \frac{2\pi}{\omega}$. Assume a truncated Fourier ansatz for the periodic response as shown in Equation 3.63.

$$z_h^{N_h}(t) = \sum_{n=0}^{N_h} z_{sn}\sin(n\omega t) + z_{cn}\cos(n\omega t) \tag{A.26}$$

Insert this into the governing differential equation and define the residual

$$R(t) \;=\; \mathrm{LHS}\left(z(t)\right) - \mathrm{RHS}. \tag{A.27}$$

Since any periodic curve can be built from a cosine part and a sine part at each harmonic, the residual takes the form

$$R(t) \;=\; c_0 + \sum_{n=1}^{N} C_n\cos(n\omega t) + D_n\sin(n\omega t), \tag{A.28}$$

where the residual coefficients $c_0, C_n, D_n$ are (generally nonlinear) functions of the unknowns $z_{c0}, \{z_{sn}, z_{cn}\}_{n=1}^{N}$ and parameters.

Using orthogonality, the residual coefficients are obtained by projection:

$$c_0 = \frac{1}{T}\int_0^T R(t)\,dt, \tag{A.29}$$

$$C_m = \frac{2}{T}\int_0^T R(t)\cos(m\omega t)\,dt, \qquad m = 1,\dots,N, \tag{A.30}$$

$$D_m = \frac{2}{T}\int_0^T R(t)\sin(m\omega t)\,dt, \qquad m = 1,\dots,N. \tag{A.31}$$

Harmonic balance enforces $R(t) \equiv 0$, which is equivalent to the algebraic system

$$c_0 = 0, \qquad C_m = 0, \qquad D_m = 0 \qquad (m = 1,\dots,N). \tag{A.32}$$

Solving these $2N+1$ equations yields the unknown Fourier coefficients $z_{c0}, \{z_{sn}, z_{cn}\}$.

Here, the equations are obtained by inserting the ansatz 3.63 into Equation 3.62 and using the gate actuation signal:

$$V_{GD}(t) = V_{GD}^{DC} + V_{GD}^{AC}\cos(\omega t) \tag{A.33}$$

The system of equations that is solved by the nonlinear solver in python `fsolve` in python are then obtained by the coefficient comparison (Eqn. A.32). The coefficient $c_0$ is static component, while $C_m$ and $D_m$ are dynamic.

## A.3.1 First Order HB Ansatz

The first order ansatz uses:

$$z_h^1(t) = z_{c0} + z_{s1}\sin(\omega t) + z_{c1}\cos(\omega t) \tag{A.34}$$

The static residual coefficient $c_0$ yields:

$$\begin{aligned} 0 &= k_0 z_{c0} + 3k_3\left(z_{c1}^2 z_{c0}/2 + z_{s1}^2 z_{c0}/2 + z_{c0}^3/3\right) \\ &+T_{ax} z_{c0}\int_0^l \left(\frac{\partial\phi}{\partial x}\right)^2 dx - \bar{C}_g'\left(V_{GD}^{AC\,2}/4 + V_{GD}^{DC\,2}/2\right)\int_0^l \phi(x)^2 dx \\ &-\bar{C}_g''\left(V_{GD}^{DC\,2} z_{c0} + V_{GD}^{AC} V_{GD}^{DC} z_{c1} + V_{GD}^{AC\,2} z_{c0}/2\right)\int_0^l \phi(x)^3 dx \end{aligned} \tag{A.35}$$

The residual coefficient $C_1$ (in front of $\cos(\omega t)$) yields:

$$
\begin{aligned}
0 = -m\omega^2 z_{c1} + b\omega z_{s1} + k_0 z_{c1} + 3k_3\left(z_{c1}^3/4 + z_{c1}z_{s1}^2/4 + z_{c1}z_{c0}^2\right) \\
-\bar{C}_g' \int_0^l \phi(x)^2 dx V_{GD}^{AC} V_{GD}^{DC} + T_{ax} z_{c1} \int_0^l \left(\frac{\partial \phi}{\partial x}\right)^2 dx \\
-\bar{C}_g'' \left(V_{GD}^{DC2} z_{c1} + 3V_{GD}^{AC2} z_{c1}/4 + 2V_{GD}^{AC} V_{GD}^{DC} z_{c0}\right) \int_0^l \phi(x)^3 dx \\
+1/4\eta\omega\left(z_{c1}^2 z_{s1} + z_{s1}^3 + 4z_{s1}z_{c0}^2\right)
\end{aligned} \quad (A.36)
$$

The residual coefficient $D_1$ (in front of $\sin(\omega t)$) yields:

$$
\begin{aligned}
0 = -m\omega^2 z_{s1} - b\omega z_{c1} + k_0 z_{s1} + T_{ax} z_{s1} \int_0^l \left(\frac{\partial \phi}{\partial x}\right)^2 dx \\
+3k_3\left(z_{s1}^3/4 + z_{s1}z_{c0}^2 + z_{c1}^2 z_{s1}/4\right) \\
-\bar{C}_g'' \left(V_{GD}^{AC2} z_{s1}/4 + V_{GD}^{DC2} z_{s1}\right) \int_0^l \phi(x)^3 dx \\
-1/4\eta\omega\left(z_{c1}z_{s1}^2 + z_{c1}^3 + 4z_{c1}z_{c0}^2\right)
\end{aligned} \quad (A.37)
$$

## A.3.2 Second Order HB Ansatz

The second order ansatz uses:

$$
z_h^2(t) = z_{c0} + z_{s1}\sin(\omega t) + z_{c1}\cos(\omega t) + z_{s2}\sin(2\omega t) + z_{c2}\cos(2\omega t) \quad (A.38)
$$

The static residual coefficient $c_0$ yields:

$$
\begin{aligned}
0 = -\bar{C}_g' \left(1/4V_{GD}^{AC2} + 1/2V_{GD}^{DC2}\right) \int_0^l \phi(x)^2 dx + T_{ax} z_{c0} \int_0^l \left(\frac{\partial \phi}{\partial x}\right)^2 dx + k_0 z_{c0} \\
+3k_3\left(z_{c1}^2 z_{c2}/4 - z_{c2}z_{s1}^2/4 + z_{c1}z_{s1}z_{s2}/2 + z_{c1}^2 z_{c0}/2 + z_{c2}^2 z_{c0}/2\right. \\
\left. + z_{s1}^2 z_{c0}/2 + z_{s2}^2 z_{c0}/2 + z_{c0}^3/3\right) \\
-\bar{C}_g'' \int_0^l \phi(x)^3 dx \left(V_{GD}^{AC} V_{GD}^{DC} z_{c1} + 1/4V_{GD}^{AC2} z_{c2} + 1/2V_{GD}^{AC2} z_{c0} + V_{GD}^{DC2} z_{c0}\right)
\end{aligned} \quad (A.39)
$$

The residual coefficient $C_1$ (in front of $\cos(\omega t)$) yields:

$$
\begin{aligned}
0 = -m\omega^2 z_{c1} + b\omega z_{s1} + k_0 z_{c1} + T_{ax} z_{c1} \int_0^l \left(\frac{\partial\phi}{\partial x}\right)^2 dx \\
+3k_3 \left(z_{c1}^3/4 + z_{c1}z_{c2}^2/2 + z_{c1}z_{s1}^2/4 + z_{c1}z_{s2}^2/2 + z_{c1}z_{c2}z_{c0} + z_{s1}z_{s2}z_{c0} + z_{c1}z_{c0}^2\right) \\
-\bar{C}_g' V_{GD}^{AC} V_{GD}^{DC} \int_0^l \phi(x)^2 dx \\
-\bar{C}_g'' \left(3/4V_{GD}^{AC2} z_{c1} + V_{GD}^{DC2} z_{c1} + 2V_{GD}^{AC} V_{GD}^{DC} z_{c0} + V_{GD}^{AC} V_{GD}^{DC} z_{c2}\right) \int_0^l \phi(x)^3 dx \\
+\eta\omega \left(z_{c1}^2 z_{s1}/4 + z_{c2}^2 z_{s1}/2 + z_{s1}^3/4 + z_{s1}z_{s2}^2/2 - z_{c2}z_{s1}z_{c0} + z_{c1}z_{s2}z_{c0} + z_{s1}z_{c0}^2\right)
\end{aligned}
\tag{A.40}
$$

The residual coefficient $D_1$ (in front of $\sin(\omega t)$) yields:

$$
\begin{aligned}
0 = -b\omega z_{c1} + k_0 z_{s1} - m\omega^2 z_{s1} + T_{ax} z_{s1} \int_0^l \left(\frac{\partial\phi}{\partial x}\right)^2 dx \\
+3k_3 \left(z_{c1}^2 z_{s1}/4 + z_{c2}^2 z_{s1}/2 + z_{s1}^3/4 + z_{s1}z_{s2}^2/2 - z_{c2}z_{s1}z_{c0} + z_{c1}z_{s2}z_{c0} + z_{s1}z_{c0}^2\right) \\
-\bar{C}_g'' \left(V_{GD}^{AC2} z_{s1}/4 + V_{GD}^{AC} V_{GD}^{DC} z_{s2} + V_{GD}^{DC2} z_{s1}\right) \int_0^l \phi(x)^3 dx \\
-\eta\omega \left(z_{c1}^3/4 + z_{c1}z_{c2}^2/2 + z_{c1}z_{s1}^2/4 + z_{c1}z_{s2}^2/2 + z_{c1}z_{c2}z_{c0} + z_{s1}z_{s2}z_{c0} + z_{c1}z_{c0}^2\right)
\end{aligned}
\tag{A.41}
$$

The residual coefficient $C_2$ (in front of $\cos(2\omega t)$) yields:

$$
\begin{aligned}
0 = -4m\omega^2 z_{c2} + 2b\omega z_{s2} + k_0 z_{c2} + T_{ax} z_{c2} \int_0^l \left(\frac{\partial\phi}{\partial x}\right)^2 dx \\
-\bar{C}_g' V_{GD}^{AC2}/4 \int_0^l \phi(x)^2 dx \\
+3k_3 \left(z_{c1}^2 z_{c2}/2 + z_{c2}^3/4 + z_{c2}z_{s1}^2/2 + z_{c2}z_{s2}^2/4 + z_{c1}^2 z_{c0}/2 - z_{s1}^2 z_{c0}/2 + z_{c2}z_{c0}^2\right) \\
-\bar{C}_g'' \left(V_{GD}^{AC2} z_{c0}/2 + V_{GD}^{AC} V_{GD}^{DC} z_{c1} + V_{GD}^{AC2} z_{c2}/2 + V_{GD}^{DC2} z_{c2}\right) \int_0^l \phi(x)^3 dx \\
+\eta\omega \left(z_{c1}^2 z_{s2} + 1/2 z_{c2}^2 z_{s2} + z_{s1}^2 z_{s2} + 1/2 z_{s2}^3 + 2z_{c1}z_{s1}z_{c0} + 2z_{s2}z_{c0}^2\right)
\end{aligned}
\tag{A.42}
$$

The residual coefficient $D_2$ (in front of $\sin(2\omega t)$) yields:

$$
\begin{aligned}
0 = -4m\omega^2 z_{s2} - 2b\omega z_{c2} + k_0 z_{s2} + T_{ax} z_{s2} \int_0^l \left(\frac{\partial\phi}{\partial x}\right)^2 dx \\
+3k_3 \left(z_{c1}^2 z_{s2}/2 + z_{c2}^2 z_{s2}/4 + z_{s1}^2 z_{s2}/2 + z_{s2}^3/4 + z_{c1}z_{s1}z_{c0} + z_{s2}z_{c0}^2\right) \\
-\bar{C}_g'' \left(V_{GD}^{AC2} z_{s2}/2 + V_{GD}^{DC2} z_{s2} + V_{GD}^{AC} V_{GD}^{DC} z_{s1}\right) \int_0^l \phi(x)^3 dx \\
-\eta\omega \left(z_{c1}^2 z_{c2} + z_{c2}^3/2 + z_{c2}z_{s1}^2 + z_{c2}z_{s2}^2/2 + z_{c1}^2 z_{c0} + 2z_{c2}z_{c0}^2 - z_{s1}^2 z_{c0}\right)
\end{aligned}
\tag{A.43}
$$

# B Current Model: Supplementary Material

## B.1 Conductive Current Component at the Lock-in Amplifier

The electrostatic gate actuation signal is described by:

$$V_{GD} = V_{GD}^{DC} + V_{GD}^{AC}\cos(\omega t) \tag{B.1}$$

The conductive current is (see Section 5.3.1):

$$i_g = \frac{dG^{DC}}{dV_g^{DC}} \frac{V_{SD}(t)}{C_0} \cdot \left( C_0 V_{GD}^{AC} \cdot \cos(\omega t) + V_g^{DC} \bar{C}_g' \int_0^l \phi dx \cdot z_d(t) \right) \tag{B.2}$$

Here, we only include the first 2 harmonics of the dynamic amplitude $z_d(t)$.

$$z_d(t) = z_{d1}\cos(\omega t) + z_{d2}\cos(2\omega t) \tag{B.3}$$

### B.1.1 $1\omega$ Readout-Technique

Read-out signal for the $1\omega$ technique is:

$$V_{SD}(t) = V_{SD}^{DC} + V_{SD}^{AC}\cos(\omega t + \Delta\omega t) \tag{B.4}$$

Combining the actuation signal (Eqn. B.1), the conductive current component (Eqn. B.2), and the $1\omega$ read out signal (Eqn. B.4), then expanding, yields:

$$i_g^\omega = \frac{dG^{DC}}{dV_g^{DC}}\frac{1}{2C_0}\cdot\left[\left(C_0V_{GD}^{AC}V_{SD}^{AC} + \bar{C}_g'V_{GD}^{DC}V_{SD}^{AC}z_{d1}\int_0^l \phi dx\right)\cos(\Delta\omega t)\right. \quad \text{(B.5)}$$

$$+\left(2C_0V_{GD}^{AC}V_{SD}^{DC} + 2\bar{C}_g'V_{GD}^{DC}V_{SD}^{DC}z_{d1}\int_0^l \phi dx\right)\cos(\omega t) \quad \text{(B.6)}$$

$$+2\bar{C}_g'V_{GD}^{DC}V_{SD}^{DC}z_{d2}\int_0^l \phi dx\cos(2\omega t) \quad \text{(B.7)}$$

$$+\bar{C}_g'V_{GD}^{DC}V_{SD}^{AC}z_{d2}\int_0^l \phi dx\cos(\Delta\omega t - \omega t) \quad \text{(B.8)}$$

$$+\left(C_0V_{GD}^{AC}V_{SD}^{AC} + \bar{C}_g'V_{GD}^{DC}V_{SD}^{AC}z_{d1}\int_0^l \phi dx\right)\cos(\Delta\omega t + 2\omega t) \quad \text{(B.9)}$$

$$\left.+\bar{C}_g'V_{GD}^{DC}V_{SD}^{AC}z_{d2}\int_0^l \phi dx\cos(\Delta\omega t + 3\omega t)\right] \quad \text{(B.10)}$$

The current component at $\Delta\omega$ is measured by the lock-in amplifier:

$$i_g^\omega = \underbrace{\frac{dG^{DC}}{2\cdot dV_{GD}^{DC}}\cdot V_{SD}^{AC}V_{GD}^{AC}}_{i_{\text{back}}} + \underbrace{\frac{dG^{DC}}{dV_{GD}^{DC}}\frac{V_{SD}^{AC}V_{GD}^{DC}}{2C_0}\cdot\bar{C}_g'\int_0^l \phi dx\cdot z_{d1}}_{i_{motional}} \quad \text{(B.11)}$$

### B.1.2 2$\omega$ Readout-Technique

For the 2$\omega$ technique, the read-out signal is:

$$V_{SD}(t) = V_{SD}^{DC} + V_{SD}^{AC}\cos(2\omega t + \Delta\omega t) \quad \text{(B.12)}$$

Combining the actuation signal (Eqn. B.1), the conductive current component (Eqn. B.2), and the 2$\omega$ read out signal (Eqn. B.12), then expanding, yields:

$$i_g = \frac{dG^{DC}}{dV_{GD}^{DC}}\frac{1}{2C_0}\cdot\left[\bar{C}_g'V_{GD}^{DC}V_{SD}^{AC}z_{d2}\int_0^l \phi dx\cos(\Delta\omega t)\right. \quad \text{(B.13)}$$

$$+\left(2C_0V_{GD}^{AC}V_{SD}^{DC} + 2\bar{C}_g'V_{GD}^{DC}V_{SD}^{DC}z_{d1}\int_0^l \phi dx\right)\cos(\omega t) \quad \text{(B.14)}$$

$$+2\bar{C}_g'V_{GD}^{DC}V_{SD}^{DC}z_{d2}\int_0^l \phi dx\cos(2\omega t) \quad \text{(B.15)}$$

$$+\left(C_0V_{GD}^{AC}V_{SD}^{AC}\int_0^l \phi dx + \bar{C}_g'V_{GD}^{DC}V_{SD}^{AC}z_{d1}\int_0^l \phi dx\right)\cos(\Delta\omega t + \omega t) \quad \text{(B.16)}$$

$$+\left(C_0V_{GD}^{AC}V_{SD}^{AC} + \bar{C}_g'V_{GD}^{DC}V_{SD}^{AC}z_{d1}\int_0^l \phi dx\right)\cos(\Delta\omega t + 3\omega t) \quad \text{(B.17)}$$

$$\left.+\bar{C}_g'V_{GD}^{DC}V_{SD}^{AC}z_{d2}\int_0^l \phi dx\cos(\Delta\omega t + 4\omega t)\right] \quad \text{(B.18)}$$

The current component at $\Delta\omega$ is measured by the lock-in amplifier:

$$i_g^{2\omega} = \frac{dG^{DC}}{dV_{GD}^{DC}} \frac{1}{2C_0} \cdot V_{SD}^{AC} V_{GD}^{DC} \bar{C}'_g \int_0^l \phi dx \cdot z_{d2} \tag{B.19}$$

# B.2 Piezo Current Component at the Lock-in Amplifier

The electrostatic gate actuation signal is described by:

$$V_{GD} = V_{GD}^{DC} + V_{GD}^{AC} \cos(\omega t) \tag{B.20}$$

The piezoresistive current is (see Section 5.3.2):

$$i_p = \frac{V_{SD}(t)}{R_0} \left( 1 - G_f \cdot \frac{z(t)^2}{2L} \int_0^L \left( \frac{\partial \phi}{\partial x} \right)^2 dx \right) \tag{B.21}$$

## B.2.1 $1\omega$ Readout-Technique

The read-out signal for $1\omega$ is:

$$V_{SD} = V_{SD}^{DC} + V_{SD}^{AC} \cos(\omega t + \Delta\omega t) \tag{B.22}$$

Combining the actuation signal (Eqn. B.1), the conductive current component (Eqn. B.21), and the $1\omega$ read out signal (Eqn. B.22), then expanding, yields:

$$\left(G_f \int_0^L \left(\frac{\partial\phi}{\partial x}\right)^2 dx/R_0/l\right) \cdot \left[V_{SD}^{DC} l/G_f/\int_0^L \left(\frac{\partial\phi}{\partial x}\right)^2 dx \quad \text{(B.23)}\right.$$

$$-V_{SD}^{DC} z_{d1}^2/4 - V_{SD}^{DC} z_{d2}^2/4 - V_{SD}^{DC} z_s^2/2 \quad \text{(B.24)}$$

$$-\left(V_{SD}^{AC} z_{d1} z_{d2}/4 - V_{SD}^{AC} z_{d1} z_s/2\right) \cos(\Delta\omega t) \quad \text{(B.25)}$$

$$-\left(V_{SD}^{DC} z_{d1} z_{d2}/2 - V_{SD}^{DC} z_{d1} z_s\right) \cos(\omega t) \quad \text{(B.26)}$$

$$-\left(V_{SD}^{DC} z_{d1}^2/4 - V_{SD}^{DC} z_{d2} z_s\right) \cos(2\omega t) \quad \text{(B.27)}$$

$$-V_{SD}^{DC} z_{d1} z_{d2}/2 \cdot \cos(3\omega t) \quad \text{(B.28)}$$

$$-V_{SD}^{DC} z_{d2}^2/4 \cdot \cos(4\omega t) \quad \text{(B.29)}$$

$$-V_{SD}^{AC} z_{d2}^2/8 \cdot \cos(\Delta\omega t - 3\omega t) \quad \text{(B.30)}$$

$$-V_{SD}^{AC} z_{d1} z_{d2}/4 \cdot \cos(\Delta\omega t - 2\omega t) \quad \text{(B.31)}$$

$$-\left(V_{SD}^{AC} z_{d1}^2/8 - V_{SD}^{AC} z_{d2} z_s/2\right) \cos(\Delta\omega t - \omega t) \quad \text{(B.32)}$$

$$+V_{SD}^{AC} l/G_f/\int_0^L \left(\frac{\partial\phi}{\partial x}\right)^2 dx - V_{SD}^{AC}\left(zd1^2/4 - z_{d2}^2/4 - z_s^2/2\right) \cos(\Delta\omega t + \omega t) \quad \text{(B.33)}$$

$$-\left(V_{SD}^{AC} z_{d1} z_{d2}/4 - (V_{SD}^{AC} z_{d1} z_s/2\right) \cos(\Delta\omega t + 2\omega t) \quad \text{(B.34)}$$

$$-\left(V_{SD}^{AC} z_{d1}^2/8 - (V_{SD}^{AC} z_{d2} z_s/2\right) \cos(\Delta\omega t + 3\omega t) \quad \text{(B.35)}$$

$$-V_{SD}^{AC} z_{d1} z_{d2}/4 \cdot \cos(\Delta\omega t + 4\omega t) \quad \text{(B.36)}$$

$$\left.-V_{SD}^{AC} z_{d2}^2/8 \cdot \cos(\Delta\omega t + 5\omega t)\right] \quad \text{(B.37)}$$

The current component at $\Delta\omega$ is measured by the lock-in amplifier:

$$i_p^{\omega} = -\frac{V_{SD}^{AC} G_f}{R_0 l} \cdot \int_0^L \left(\frac{\partial\phi}{\partial x}\right)^2 dx \cdot \left(\frac{z_{d1} z_s}{2} + \frac{z_{d1} z_{d2}}{4}\right) \tag{B.38}$$

### B.2.2 2$\omega$ Readout-Technique

The read-out signal for $2\omega$ is:

$$V_{SD} = V_{SD}^{DC} + V_{SD}^{AC} \cos\left(2\omega t + \Delta\omega t\right) \tag{B.39}$$

Combining the actuation signal (Eqn. B.1), the conductive current component (Eqn. B.21), and the $1\omega$ read out signal (Eqn. B.39), then expanding, yields:

$$\left(\frac{G_f}{R_0 l}\int_0^L\left(\frac{\partial\phi}{\partial x}\right)^2 dx\right)\cdot\Big[V_{SD}^{DC}l-(V_{SD}^{DC}z_{d1}^2)/4-V_{SD}^{DC}z_{d2}^2/4-V_{SD}^{DC}z_s^2/2 \tag{B.40}$$

$$-(V_{SD}^{AC}z_{d1}z_{d2}/4-V_{SD}^{AC}z_{d1}z_s/2)\cos(\Delta\omega t) \tag{B.41}$$

$$-(V_{SD}^{DC}z_{d1}z_{d2}/2-V_{SD}^{DC}z_{d1}z_s)\cos(\omega t) \tag{B.42}$$

$$-(V_{SD}^{DC}z_{d1}^2/4-V_{SD}^{DC}z_{d2}z_s)\cos(2\omega t) \tag{B.43}$$

$$-V_{SD}^{DC}z_{d1}z_{d2}/2\cos(3\omega t) \tag{B.44}$$

$$-V_{SD}^{DC}z_{d2}^2/4\cos(4\omega t) \tag{B.45}$$

$$-V_{SD}^{AC}z_{d2}^2/8\cos(\Delta\omega t-3\omega t) \tag{B.46}$$

$$-V_{SD}^{AC}z_{d1}z_{d2}/4\cos(\Delta\omega t-2\omega t) \tag{B.47}$$

$$-(V_{SD}^{AC}z_{d1}^2/8-V_{SD}^{AC}z_{d2}z_s)/2)\cos(\Delta\omega t-\omega t) \tag{B.48}$$

$$+(V_{SD}^{AC}l-V_{SD}^{AC}z_{d1}^2/4-V_{SD}^{AC}z_{d2}^2/4-V_{SD}^{AC}z_s^2/2)\cos(\Delta\omega t+\omega t) \tag{B.49}$$

$$-(V_{SD}^{AC}z_{d1}z_{d2}/4-V_{SD}^{AC}z_{d1}z_s/2)\cos(t+2\omega t) \tag{B.50}$$

$$-(V_{SD}^{AC}z_{d1}^2/8-V_{SD}^{AC}z_{d2}z_s/2)\cos(t+3\omega t) \tag{B.51}$$

$$-V_{SD}^{AC}z_{d1}z_{d2}/4\cos(\Delta\omega t+4\omega t) \tag{B.52}$$

$$-\,V_{SD}^{AC}z_{d2}^2/8\cos(\Delta\omega t+5\omega t)\Big] \tag{B.53}$$

The current component at $\Delta\omega$ is measured by the lock-in amplifier:

$$i_p^{2\omega}=-\frac{V_{SD}^{AC}G_f}{R_0 l}\cdot\int_0^L\left(\frac{\partial\phi}{\partial x}\right)^2 dx\cdot\left(\frac{z_{d1}^2}{8}+\frac{z_1 z_{d2}}{2}\right) \tag{B.54}$$

# C Detailed Fabrication Procedure

## C.1 L1: SiN Etch-Through

**Table C.1:** SiN removal to reduce effect of internal layer stress

| | Step | Tool | Parameters |
|---|---|---|---|
| Local SiNx removal | Adhesion Layer | Spinner bench | HMDS |
| | Resist spinning | Spin coater | AZ1512, 4k rpm, 40 s |
| | Resist bake | Hotplate | 110°, 1 min |
| | UV exposure | MA6 | 3.6 s hard L1 Mask |
| | Develop | Spinner bench | AZ400K:DI water 1:4 |
| | Postbake | Hotplate | 1 min, 110° |
| | SiNx etch through | RIE | 23 min over etch |
| | Stripping | Wetbench | Acetone, IPA rinse |
| | Cleaning | O2 Asher | 600 W, 3 min |

## C.2 L2: Gate Patterning

**Table C.2:** SiN gate patterning

| | Step | Tool | Parameters |
|---|---|---|---|
| Gate patterning | Adhesion layer | Spinner bench | HMDS |
| | Resist spinning | Spin coater | AR-N 7520.17, 4k rpm, 1 min |
| | Resist bake | Hotplate | 85 °C, 90 s |
| | UV exposure | MA6 | 3.2 s hard L2 Mask |
| | Ebeam exposure | EBPG 5200 | Service eL3 |
| | Develop | Wetbench | AZ300-47 pure, 4 min |
| | Postbake | Hotplate | 110 °C, 1 min |
| | SiNx etching | RIE | 14 min |
| | Stripping | RIE | O2 plasma, 10 min |

## C.3 L3: Electrode Deposition

**Table C.3:** First step of electrode deposition

| | Step | Tool | Parameters |
|---|---|---|---|
| Ebeam electrodes | Adhesion layer | Spinner bench | HMDS |
| | Resist spinning | Spin coater | AR-P 617.06, 6k rpm, 60 s |
| | Resist bake | Hot plate | 180 °C, 5 min |
| | Resist spinning | Spinner bench | AR-P 672.02, 4k rpm, 60 s |
| | Resist bake | Hot plate | 180 °C, 5 min |
| | E-beam exposure | EBPG 5200 | Service Layer 1 |
| | Develop | Wetbench | MIBK:IPA 1:2, 150 s |
| | Rinse | Wetbench | IPA rinse, 30 s |
| | Metal evaporation | Evatec | 2 nm Cr, 100 nm Pd |
| | Lift-off | Wetbench | Acetone, USB until finished |
| | Stripping | O2 Asher | 600 W, 3 min |
| UV electrodes | Adhesion | Hotplate | 180 °C, 5 min |
| | Resist spinning | Spin coater | LOR 5B, 4k rpm, 40 s |
| | Resist bake | Hotplate | 180 °C, 5 min |
| | Resist spinning | Spin coater | AZ 1512, 4k rpm, 40 s |
| | Resist bake | Hotplate | 110°, 1 min |
| | UV exposure | MA-6 | Vac. mode, 3.6 s L3 Mask |
| | Develop | Wetbench | DI-H2O:AZ400k 4:1 |
| | Metal evaporation | Evatec | 2 nm Cr, 100 nm Pd |
| | Stripping | Wetbench | DMSO, 130 °C, 1 h |

## C.4 L4: Beam Shape Patterning

**Table C.4:** Beam shape patterning

| | Step | Tool | Parameters |
|---|---|---|---|
| Beam shape patterning | Adhesion layer | Spinner bench | HMDS |
| | Resist spinning | Spin coater | AR-N 7520.17, 1k rpm, 60 s |
| | Resist bake | Hotplate | 85 °C, 6 min |
| | UV exposure | MA6 | Vacuum, X s, L4 Mask |
| | E-beam exposure | EBPG 5200 | Service eL6 |
| | Develop | Wetbench | AZ 300-47, pure, 11 min |
| | SiN removal | RIE | 22 min over etch |
| | Si etch | DRIE | 100 cycles, movo_0C |
| | Stripping | RIE | plasma removal, 10 min |

## C.5 L5: Mesa Etch

**Table C.5:** Mesa etch step

| | Step | Tool | Parameters |
|---|---|---|---|
| Mesa etch | Dehydration | Hotplate | 180 °C, 5 min |
| | Adhesion layer | Spinner bench | HMDS |
| | Resist spinning | Spin coater | AZ 4562, 4k rpm, 40 s |
| | Resist bake | Hotplate | 110 °C, 60 s |
| | UV exposure | MA6 | hard, 14.5 s L5 Mask |
| | Development | Wetbench | DI-H2O:AZ400K 3:1, 50 s |
| | Hard bake | Hotplate | 110 °C, 3 min no shock cooling |
| | Oxide removal | HF wetbench | BHF 1:7, 24.5 min |
| | Si etch | DRIE | 0_ASE_10C, 30 cycles |
| | Stripping | Wetbench | DMSO, 130 °C, 60 min |

## C.6 L6: Backside Etch and Release

**Table C.6:** Backside etch and release

| | Step | Tool | Parameters |
|---|---|---|---|
| SiN removal | Adhesion layer | Spinner bench | HMDS |
| | Resist spinning | Spin coater | AZ4533, 4k rpm, 40 s |
| | Resist bake | Hotplate | 110 °C, 1 min |
| | SiN removal | RIE | 25 min |
| | Stripping | Hot plate | DMSO, 130 °C, 60 min |
| Backside etch and release | Adhesion layer | Spinner bench | HMDS |
| | Resist spinning | Spin coater | AZ 4562, 2k rpm, 40 s |
| | Resist bake | Hotplate | 110 °C, 10 min |
| | UV exposure | MA8 | Hard contact, 18 s L6 Mask |
| | Development | Wetbench | DI-H2O:AZ400K 1:3 |
| | Put on carrier | Hotplate | with wax pallets |
| | Backside etch | DRIE | ASE_10_C, 450 cycles |
| | Stripping | Wetbench | DMSO, 130 °C, 1.5 h |
| | Stripping | Plasma asher | 600 W, 5 min |
| | BHF release | HF wetbench | BHF 1:7, 24 min |

## C.7 Marker Design

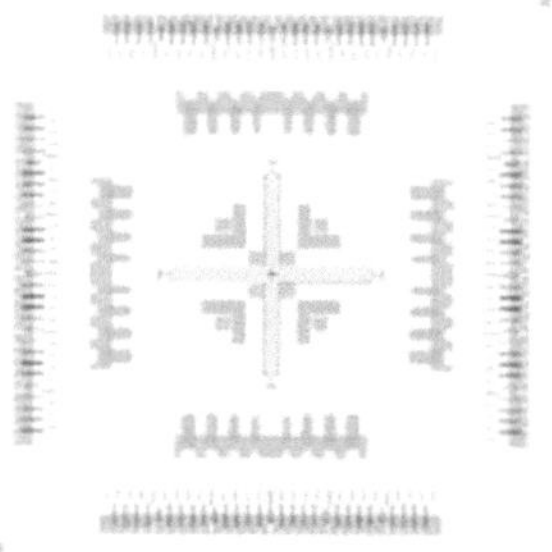

(a) Alignment marker.

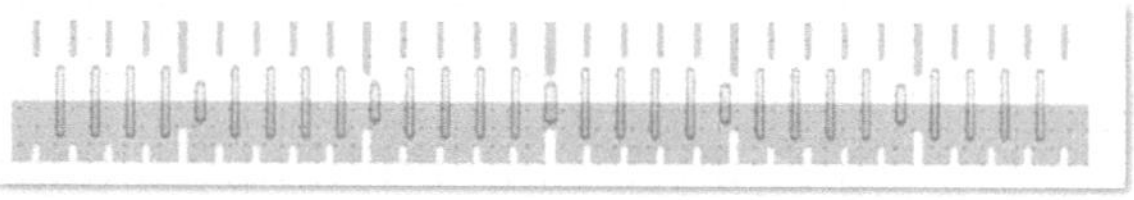

(b) Vernier Scale with a different spacing than the reference to estimate the missalignment.

**Figure C.1:** Alignment marker used to increase the precision of alignment. The central cross is the marker deposited in L0.

# C.8 Datasheet: Wafer

**Icemos Technology Ltd** **Product Specification** **1000.567002** Issue Date 12 October 2021 08:3

Part Number ETH SOI 001 REV01, Customer Part: ETH SOI 001 REV01 Customer ETH Zurich

| Category | | Parameter | Specification | Measurement Method |
|---|---|---|---|---|
| OverallWafer | 1.0 | Diameter | 100.00 +/- 0.50 mm | |
| | 2.0 | Primary Flat Orientation | <110> +/- 1 degree | Wafer Vendor |
| | 3.0 | Primary Flat Length | 32.50 +/- 2.50 mm | Wafer Vendor |
| | 4.0 | Secondary Flat Orientation | none or Semi Std | |
| | 5.0 | Overall Thickness | 407.00 +/- 11.00 µm | ADE 100% |
| | 6.0 | Total Thickness Variation (TTV) | <5.00µm | Guaranteed by Process |
| | 7.0 | Bow | <120.00µm | ADE to ASTM F534, 20% |
| | 8.0 | Warp | <120.00µm | ADE to ASTM F657, 20% |
| | 9.0 | Edge Chips | 0 | Bright Light, 100% |
| | 10.0 | Edge Exclusion | 5mm | |
| HandleSilicon | 11.0 | Handle Growth Method | CZ | Wafer Vendor |
| | 12.0 | Handle Orientation | {100} +/- 1 degree | Wafer Vendor |
| | 13.0 | Handle Thickness | 400.00 +/- 10.00 µm | ADE, 100% |
| | 14.0 | Handle Doping Type | P | Wafer Vendor |
| | 15.0 | Handle Dopant | Boron | Wafer Vendor |
| | 16.0 | Handle Resistivity | <0.02 ohm-cm | Wafer Vendor |
| | 17.0 | Backside Finish | Polished with lasermark (no oxide) | Wafer Vendor |
| BuriedOxide | 18.0 | Oxide Type | Thermal | |
| | 19.0 | Oxide Thickness | 20,000.00 +/- 1,000.00 A | Nanospec centre point, 4% |
| | 20.0 | Oxide formed on | Handle/Device Wafer | |
| DeviceSilicon | 21.0 | Device Growth Method | CZ | Wafer Vendor |
| | 22.0 | Device Orientation | {100} +/- 1 degree | Wafer Vendor |
| | 23.0 | Nominal Thickness | 5.00 +/- 0.50 µm | FTIR, 100% 9-Pt (note3) |
| | 24.0 | Distance to device silicon edge from wafer edge | < 2 mm | Guaranteed by Process |
| | 25.0 | Device Doping Type | P | Wafer Vendor |
| | 26.0 | Device Dopant | Boron | Wafer Vendor |
| | 27.0 | Device Resistivity | <0.02 ohm-cm | Wafer Vendor |
| | 29.0 | Voids | 0 | Bright Light, 100% (note 2) |
| | 30.0 | Scratches | total length <5mm | Bright Light, 100% (note 2) |
| | 31.0 | Haze | none | Bright Light, 100% (note 2) |

Page 1 of 2 26/10/2021 www.icemostech.com

# D Device Directory

Here, all devices measured in this work are summarized.

## D.1 Devices without Axial Straining Capability

In this section, the two devices, which have been analysed in chapter 5 are presented. Their device architecture differs from the presented MEMS straining platforms with thermal actuators presented in chapter4. Both followed the same fabrication approach presented in [137] and only have the potential to apply strain by gate bias.

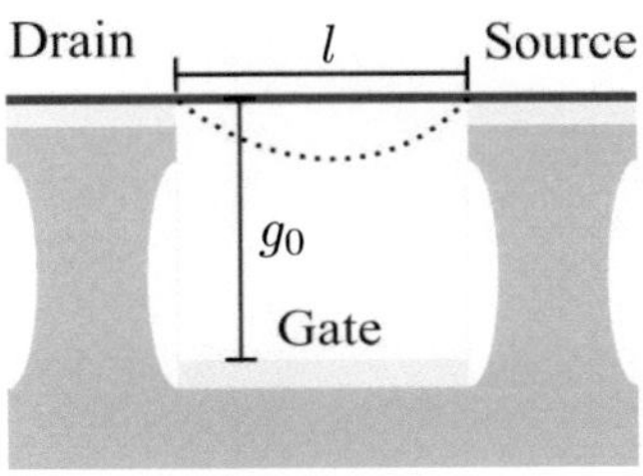


**Figure D.1:** Device schematic.

| | Device | CGDS 21 | SJDS 123 |
|---|---|---|---|
| CNT | $l$ | 2.1 µm | 2 µm |
| | $g_0$ | 2 µm | 2 µm |
| | $E$ | 1 TPa | 1 TPa |
| | $h$ | 340 pm | 340 pm |
| | CNT type | SGS | SGS |
| | $R_0$ | 25 kΩ | 62 kΩ |
| | fabricated by | C. Gentili | Dr. S. Jung |
| Fitted | $r$ | 1.1 nm | 0.8 nm |
| | $\alpha_{ML}$ | 6 | 2.3 |
| | $\epsilon_0$ | 0.03 % | 0.04 % |
| | $G_f$ | 1100 | 620 |

**Table D.1:** Device properties.

## D.2 SJDS 123

The same measurements as for device CGDS 21 have been performed on SJDS 123 yielding similar fitting parameters.

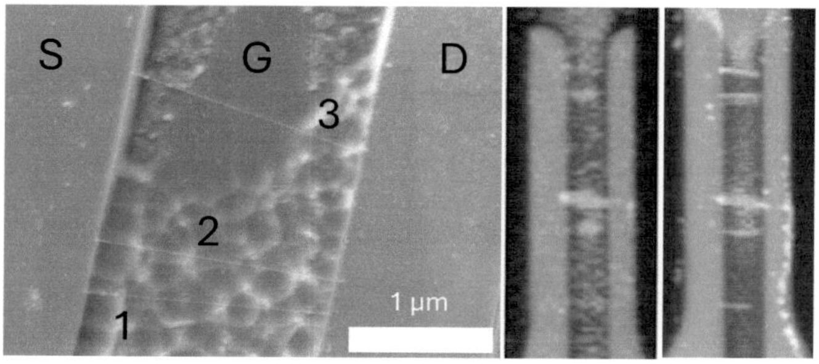


**Figure D.2:** SEM and Raman maps on device CGDS 21. SEM shows 3 CNTs, while Raman can map the full device and finds up to 6 CNTs. This can explain the low device resistance. Raman image courtesy Miroslav Haluska.

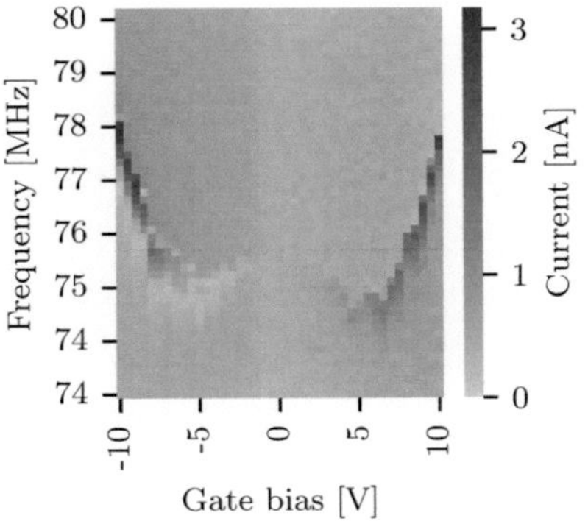


**(a)** $1\omega$ current.

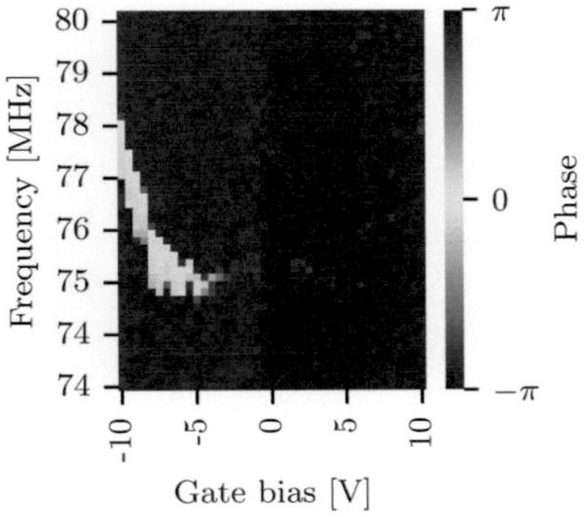


**(b)** $1\omega$ phase.

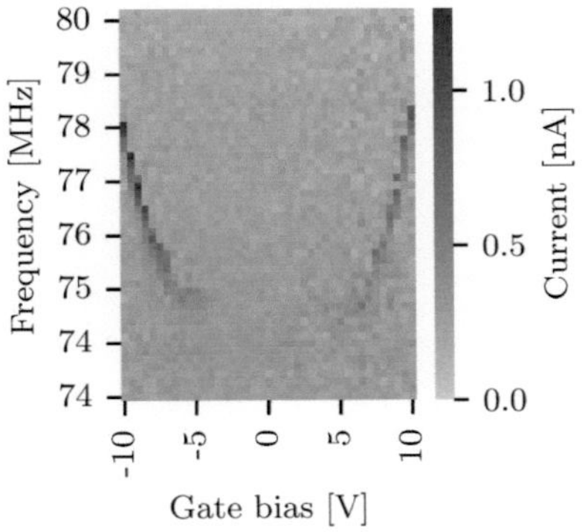


**(c)** $2\omega$ current.

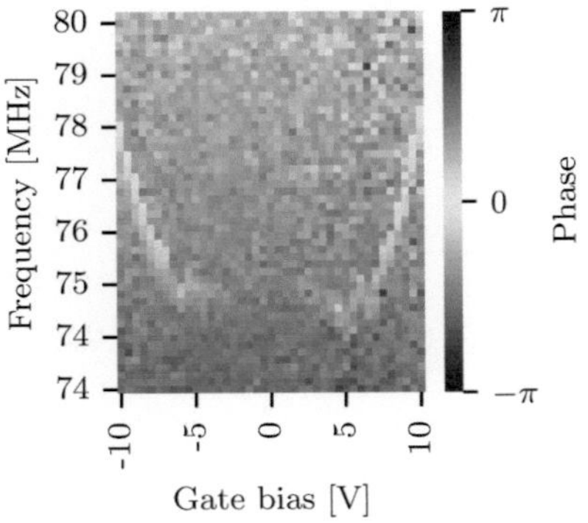


**(d)** $2\omega$ Phase.

**Figure D.3:** Device SJDS 123 full result overview showing current amplitude and phase for different gate biases and both $\omega$ read out techniques. The eigenfrequency was tuned from 73.5 MHz up to 78 MHz. $V_{GD}^{AC} = 10\,\text{mV}$, $V_{SD}^{AC} = 10\,\text{mV}$.

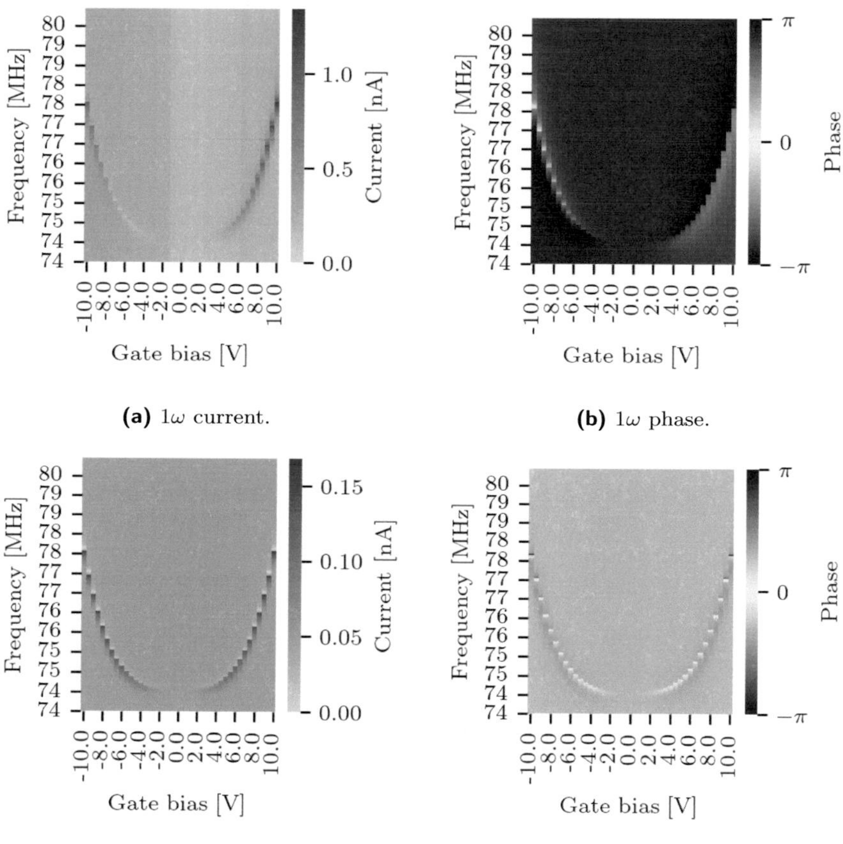


**(a)** 1ω current.

**(b)** 1ω phase.

**(c)** 2ω current.

**(d)** 2ω Phase.

**Figure D.4:** Modelled current and phase responses of device SJDS 123.

Just as device CGDS 21 the resonance current shows an asymmetry with gate bias, which was replicated by the model. It appears, that for low gate biases a spring softening effect might be present, because the frequency decreases up to 5 V gate bias $V_{GD}^{DC}$. This effect is surprising, because the spring softening effect is expected to be too weak for CNT with such capacitances.

The $1\omega$ and $2\omega$ measurement at both 10 V and −10 V gate bias are shown in figure D.5. The eigenfrequencies for 10 V are not overlapping. The different shape of the resonance peak is more visible than for CGDS 21.

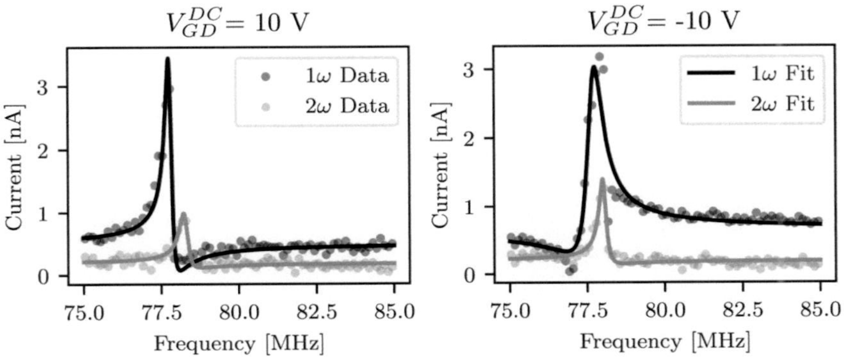


**Figure D.5:** $1\omega$ and $2\omega$ read-out for gate bias 10 V (left) and −10 V (right).

# D.3 Devices with Axial Straining Capability

All other devices presented in this work follow the design and fabrication presented in chapter 4 (schematic in Fig. D.6).

Devices used to characterise the actuator displacements are again summarized in table D.2. The image acquisition of the SEM images was performed by Lorenzo Fulciniti in a semester project. Devices intended for CNT measurements have passed the electrical resistance testing and are labelled with MV. Devices that showed good resistances for the thermal actuators, but shorts to one of the electrodes are labelled with L.

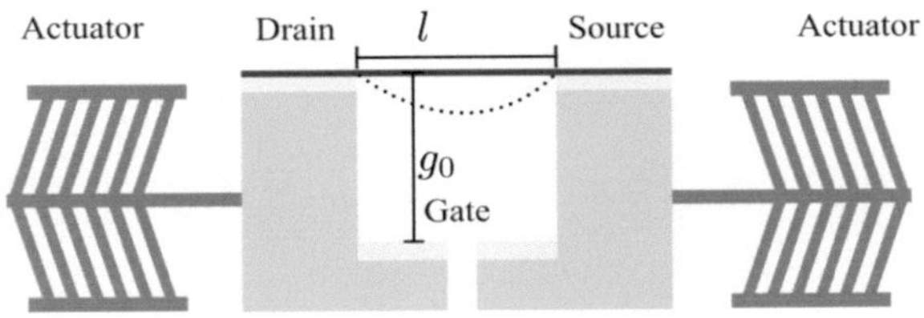


**Figure D.6:** Schematic of strainable MEMS platforms.

**Table D.2:** Fitted displacement coefficients $a$ from SEM image analysis of all devices in comparison to simulated displacement coefficients. The uncertainty in the COMSOL simulation $a$ reflects the uncertainty of the device layer thickness.

| Design | Device | Angle | Springs | SEM $a$ | COMSOL $a$ |
|---|---|---|---|---|---|
| A6 | L2 | 30 | 0 | $2.3\,\mathrm{nm\,mA^{-2}}$ | $(1.60 \pm 0.23)\,\mathrm{nm\,mA^{-2}}$ |
| | MV20 | 35 | 2 | $2.61\,\mathrm{nm\,mA^{-2}}$ | |
| A9 | L5 | 40 | 2 | $1.13\,\mathrm{nm\,mA^{-2}}$ | $(0.95 \pm 0.13)\,\mathrm{nm\,mA^{-2}}$ |
| | L6 | 40 | 2 | $1.31\,\mathrm{nm\,mA^{-2}}$ | |
| | MV12 | 40 | 2 | $0.97\,\mathrm{nm\,mA^{-2}}$ | |
| | MV21 | 40 | 2 | $1.01\,\mathrm{nm\,mA^{-2}}$ | |
| A13 | L3 | 35 | 2 | $1.03\,\mathrm{nm\,mA^{-2}}$ | $(0.85 \pm 0.12)\,\mathrm{nm\,mA^{-2}}$ |
| | MV01 | 40 | 2 | $1.03\,\mathrm{nm\,mA^{-2}}$ | |
| | L7 | 30 | 2 | $0.58\,\mathrm{nm\,mA^{-2}}$ | |

**Table D.3:** Devices for resistance characterization.

| Device | Design |
|---|---|
| E4 | A6 |
| E2 | A9 |
| E6 | A9 |
| E8 | A9 |
| E3 | A13 |

A last measurement series investigating the resistance of the thermal actuators in vacuum has been performed by Eric Meier in his Bachelor's thesis. Devices for his experiments were labelled with E and are summarized in table D.3.

## D.4 Slacked CNT on MV 01

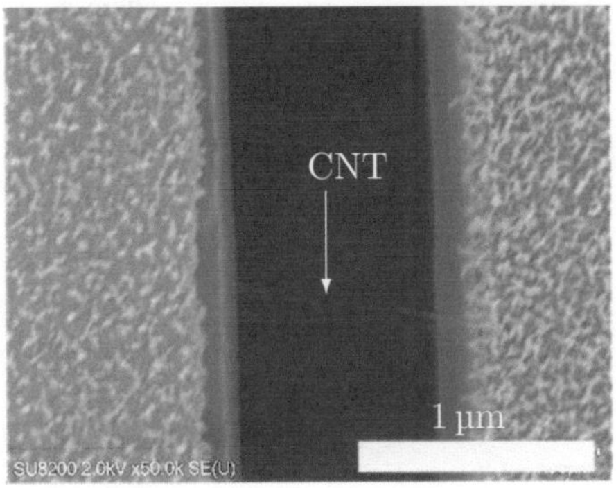


**Figure D.7:** Visible slack in the CNT of device MV 01 at $I_{act} = 0\,\mathrm{mA}$ after it had slipped at $I_{act} = 8.1\,\mathrm{mA}$. The slack is seen, because the CNT is not seen as a straight line.

## D.5 HB Fitting Results on Device MV39

Here, the HB analysis results for device MV 39 are presented as shown for device MV 12 in Section 6.3.

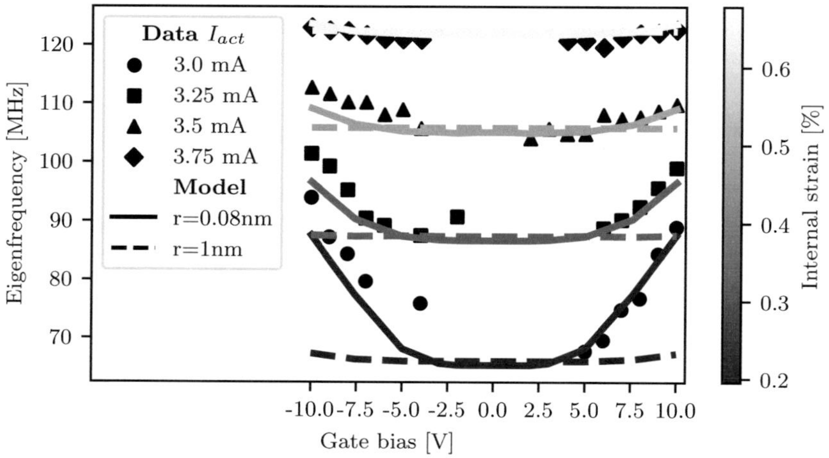


**Figure D.8:** HB fit for Device MV 39 for different radii with unchanged $E \cdot h$.

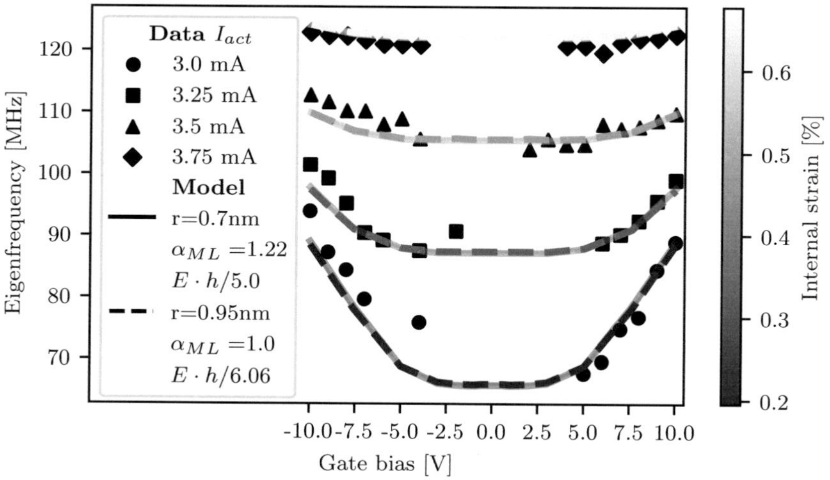


**Figure D.9:** HB fit for Device MV 39 for different radii with changed $E \cdot h$.

# E Actuator Directory

## E.1 L2

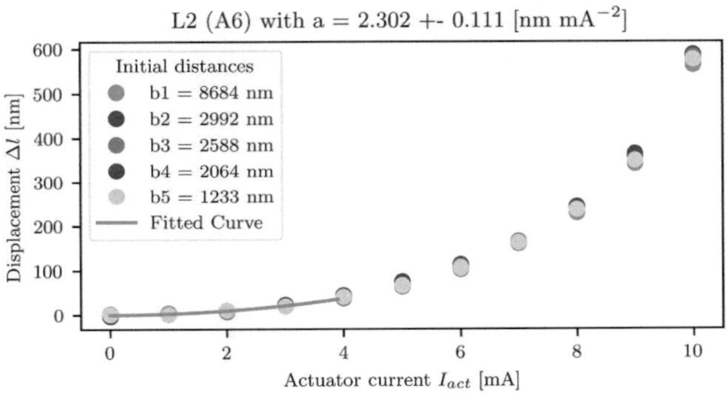


**Figure E.1:** Current-displacement characteristic of device L2.

## E.2 MV 20

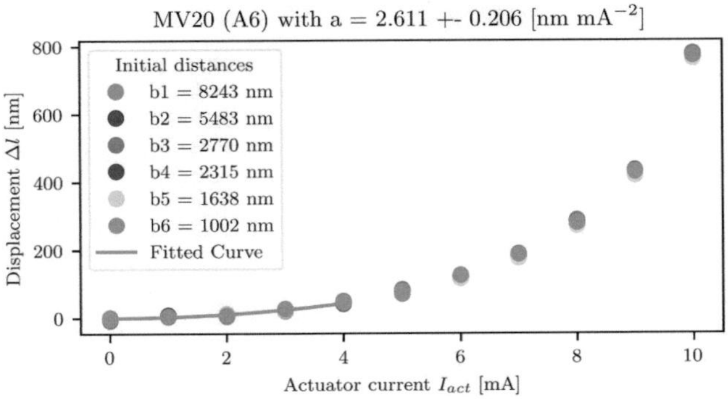


**Figure E.2:** Current-displacement characteristic of device MV20.

## E.3 L5

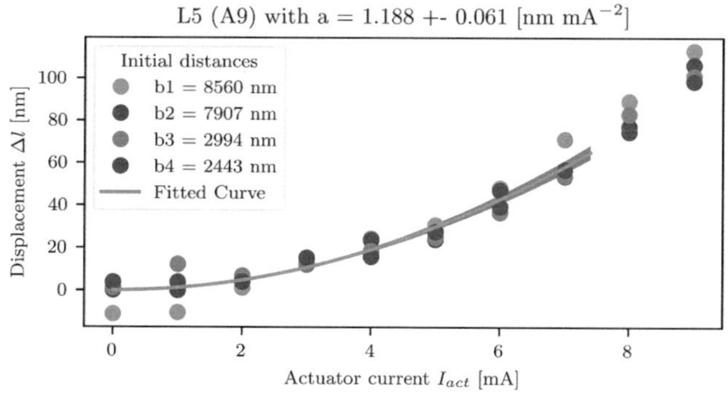


**Figure E.3:** Current-displacement characteristic of device L5.

## E.4 L6

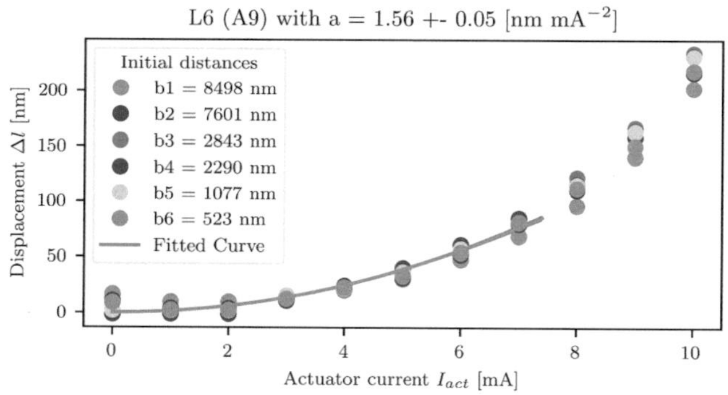


**Figure E.4:** Current-displacement characteristic of device L6.

## E.5 MV 12

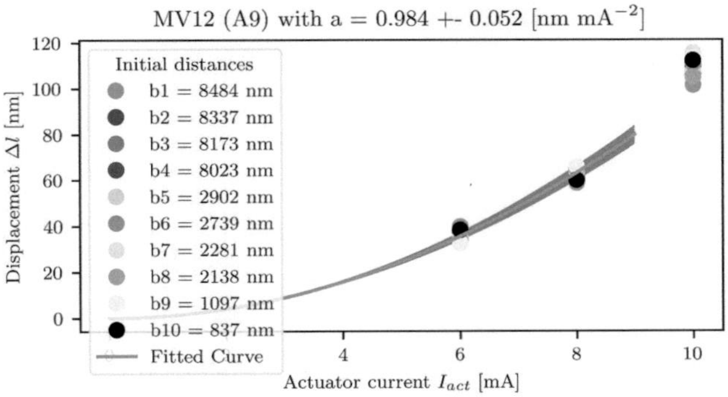


**Figure E.5:** Current-displacement characteristic of device MV 12.

## E.6 MV 21

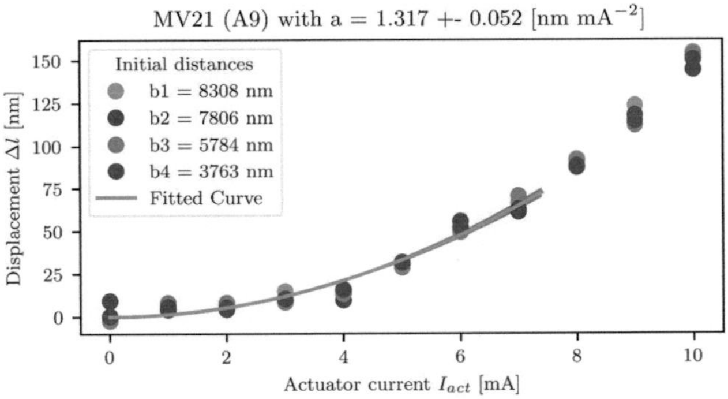


**Figure E.6:** Current-displacement characteristic of device MV 21.

## E.7 L3

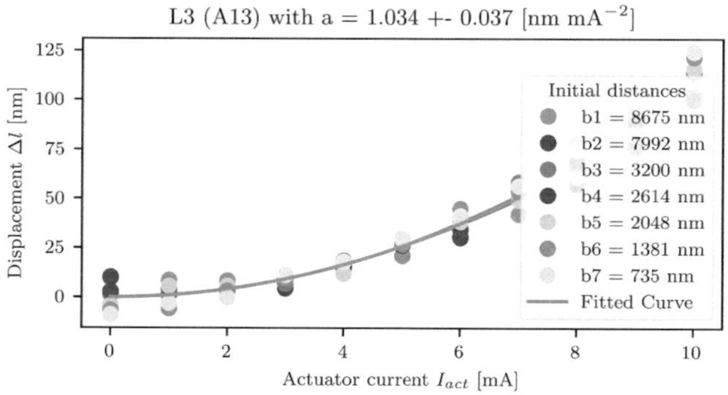


**Figure E.7:** Current-displacement characteristic of device L3.

## E.8 L7

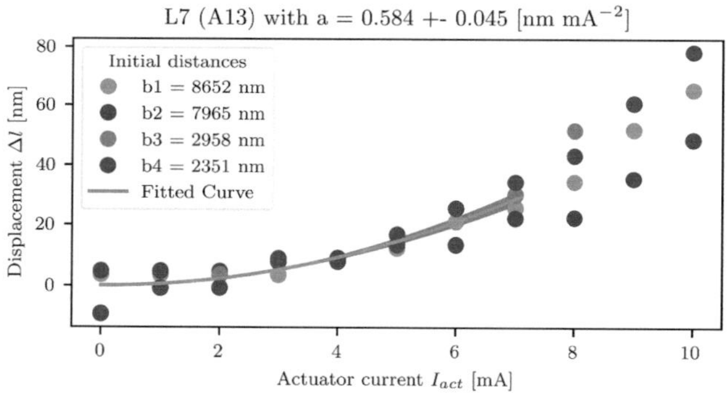


**Figure E.8:** Current-displacement characteristic of device L7.

# E.9 MV 01

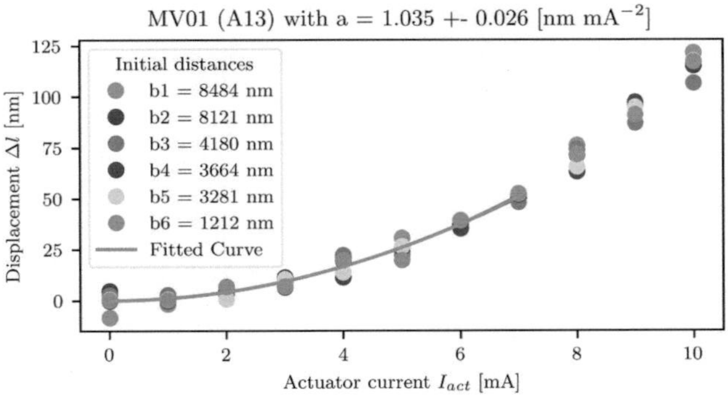


**Figure E.9:** Current-displacement characteristic of device MV 01.

# F COMSOL Simulation: Supplementary Material

The thermal conductivity $\kappa$ is adapted from the work presented in [142], where it was measured every 50 K from 100 K to 1350 K. The values are interpolated in COMSOL with a piece-wise cubic method for the final function shown in Figure F.1 as solid line.

The electrical resistivity is derived from the electrical conductance $\sigma(\mathrm{T})$ (Eqn. 4.3). In Figure F.1 shows a maximum resistivity before the the intrinsic carrier become available.

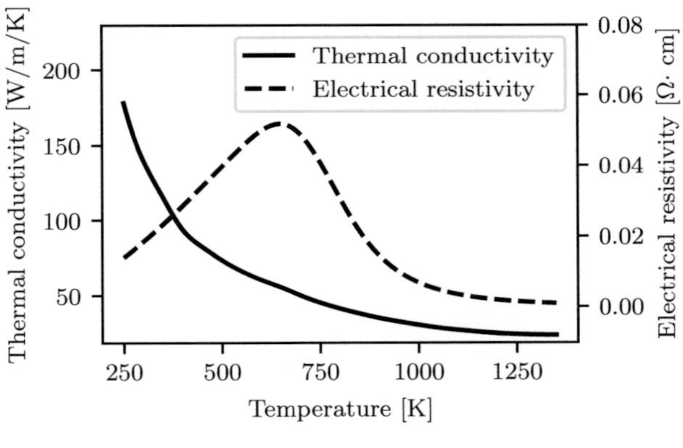


**Figure F.1:** Thermal conductivity and electrical resistivity used for simulation.

The silicon of the device layer is anisotropic and its elasticity matrix is represented by

$$\begin{bmatrix} \sigma_1 \\ \sigma_2 \\ \sigma_3 \\ \sigma_4 \\ \sigma_5 \\ \sigma_6 \end{bmatrix} = \begin{bmatrix} 194.5 & 35.7 & 64.1 & 0 & 0 & 0 \\ 35.7 & 194.5 & 64.1 & 0 & 0 & 0 \\ 64.1 & 64.1 & 165.7 & 0 & 0 & 0 \\ 0 & 0 & 0 & 79.6 & 0 & 0 \\ 0 & 0 & 0 & 0 & 79.6 & 0 \\ 0 & 0 & 0 & 0 & 0 & 50.9 \end{bmatrix} \begin{bmatrix} \mathcal{E}_1 \\ \mathcal{E}_2 \\ \mathcal{E}_3 \\ \mathcal{E}_4 \\ \mathcal{E}_5 \\ \mathcal{E}_6 \end{bmatrix}. \quad \text{(F.1)}$$

# Bibliography


[1] R. L. Zettler, *Woolley's Excavations at Ur: New Perspectives from Artifact Inventories, Field Records, and Archival Documentation.* University Park, USA: Penn State University Press, 2021, pp. 7–34, DOI: http://dx.doi.org/doi:10.1515/9781646021512-005.

[2] G. Galilei, "De motu antiquiora," in *Le Opere di Galileo Galilei, Edizione Nazionale*, A. Favaro, Ed. Firenze: G. Barbèra, 1890, vol. I, pp. 251–348, originally composed ca. 1590.

[3] W. A. Marrison, "The evolution of the quartz crystal clock," *The Bell System Technical Journal*, vol. 27, no. 3, pp. 510–588, 1948, DOI: http://dx.doi.org/10.1002/j.1538-7305.1948.tb01343.x.

[4] G. Moore, "Cramming More Components Onto Integrated Circuits," *Proceedings of the IEEE*, vol. 86, no. 1, pp. 82–85, 1998, DOI: http://dx.doi.org/10.1109/JPROC.1998.658762.

[5] H. Fujita, "A decade of MEMS and its future," in *Proceedings IEEE The Tenth Annual International Workshop on Micro Electro Mechanical Systems. An Investigation of Micro Structures, Sensors, Actuators, Machines and Robots*, 1997, pp. 1–7, DOI: http://dx.doi.org/10.1109/MEMSYS.1997.581729.

[6] K. Petersen, "Silicon as a mechanical material," *Proceedings of the IEEE*, vol. 70, no. 5, pp. 420–457, 1982, DOI: http://dx.doi.org/10.1109/PROC.1982.12331.

[7] J. L. Hilbert, "RF-MEMS for Wireless Communications," *IEEE Communications Magazine*, vol. 46, no. 8, pp. 68–74, 2008, DOI: http://dx.doi.org/10.1109/MCOM.2008.4597106.

[8] S. Iijima, "Helical microtubules of graphitic carbon," *Nature*, vol. 354, no. 6348, pp. 56–58, 1991, DOI: http://dx.doi.org/10.1038/354056a0.

[9] K. S. Novoselov, A. K. Geim, S. V. Morozov, D.-e. Jiang, Y. Zhang, S. V. Dubonos, I. V. Grigorieva, and A. A. Firsov, "Electric Field Effect in Atomically Thin Carbon Films," *Science*, vol. 306, no. 5696, pp. 666–669, 2004, DOI: http://dx.doi.org/10.1126/science.1102896.

[10] C. C. W. Ruppel, "Acoustic wave filter technology - a review," *IEEE Transactions on Ultrasonics, Ferroelectrics, and Frequency Control*, vol. 64, no. 9, pp. 1390–1400, 2017, DOI: http://dx.doi.org/10.1109/TUFFC.2017.2690905.

[11] G. Piazza, P. J. Stephanou, and A. P. Pisano, "Piezoelectric Aluminum Nitride Vibrating Contour-Mode MEMS Resonators," *Journal of Microelectromechanical Systems*, vol. 15, no. 6, pp. 1406–1418, 2006, DOI: http://dx.doi.org/10.1109/JMEMS.2006.886012.

[12] G. Wingqvist, "AlN-based sputter-deposited shear mode thin film bulk acoustic resonator (FBAR) for biosensor applications - A review," *Surface and Coatings Technology*, vol. 205, no. 5, pp. 1279–1286, 2010, DOI: http://dx.doi.org/10.1016/j.surfcoat.2010.08.109.

[13] M. Park, J. Wang, and A. Ansari, "High-overtone thin film ferroelectric AlScN-on-silicon composite resonators," *IEEE Electron Device Letters*, vol. 42, no. 6, pp. 911–914, 2021, DOI: http://dx.doi.org/10.1109/LED.2021.3070274.

[14] V. Plessky, S. Yandrapalli, P. Turner, L. Villanueva, J. Koskela, and R. Hammond, "5 GHz laterally-excited bulk-wave resonators (XBARs) based on thin platelets of lithium niobate," *Electronics Letters*, vol. 55, no. 2, pp. 98–100, 2019, DOI: http://dx.doi.org/10.1049/el.2018.7297.

[15] T.-H. Hsu, J. Campbell, J. Kramer, S. Cho, M.-H. Li, and R. Lu, "C-Band Lithium Niobate on Silicon Carbide SAW Resonator With Figure-of-Merit of 124 at 6.5 GHz," *Journal of Microelectromechanical Systems*, vol. 33, no. 5, pp. 604–609, 2024, DOI: http://dx.doi.org/10.1109/JMEMS.2024.3423768.

[16] L. Garcia-Gancedo, F. Al-Naimi, A. J. Flewitt, W. I. Milne, G. M. Ashley, J. K. Luo, X. Zhao, and J. R. Lu, "ZnO-Based FBAR resonators with carbon nanotube electrodes," *IEEE Transactions on Ultrasonics, Ferroelectrics, and Frequency Control*, vol. 58, no. 11, pp. 2438–2445, 2011, DOI: http://dx.doi.org/10.1109/TUFFC.2011.2100.

[17] N. Yamauchi, T. Shirai, T. Yoshihara, Y. Hayasaki, T. Ueda, T. Matsushima, K. Wasa, I. Kanno, and H. Kotera, "High coupling piezoelectric thin films of Pb(Zr,Ti)O3-based ternary perovskite compounds for GHz-range film bulk acoustic resonators," *Applied Physics Letters*, vol. 94, no. 17, p. 172903, 2009, DOI: http://dx.doi.org/10.1063/1.3126060.

[18] J. Masson, D. Gachon, L. Robert, N. Bazin, J. Friedt, G. Martin, S. Alzuaga, R. Rodolphe, B. Guichardaz, and S. Ballandras, "High Overtone Bulk Acoustic Resonators built using Aluminum Nitride thin films deposited onto AT-cut quartz plates," in *2006 IEEE International Frequency Control Symposium and Exposition*, 2006, pp. 835–838, DOI: http://dx.doi.org/10.1109/FREQ.2006.275498.

[19] Vinita, D. Pareek, S. Masiul Islam, and J. Singh, "Bulk acoustic wave resonators for sensing applications: A review," *Sensors and Actuators A: Physical*, vol. 378, p. 115839, 2024, DOI: http://dx.doi.org/10.1016/j.sna.2024.115839.

[20] A. Vorobiev and S. Gevorgian, "Enhanced tunable performance of high Q-factor $Ba_xSr_{1-x}TiO_3$ film bulk acoustic wave resonators," *International Journal of Microwave and Wireless Technologies*, vol. 5, no. 3, pp. 361–369, 2013, DOI: http://dx.doi.org/10.1017/S1759078713000317.

[21] C. K. Kent, N. Ramakrishnan, and H. Pratama Kesuma, "Advancements in One-Port Surface Acoustic Wave (SAW) Resonators for Sensing Applications: A Review," *IEEE Sensors Journal*, vol. 24, no. 11, pp. 17 337–17 352, 2024, DOI: http://dx.doi.org/10.1109/JSEN.2024.3386917.

[22] W. Geng, C. Zhao, F. Xue, X. Qiao, J. He, G. Xue, Y. Liu, H. Wei, K. Bi, L. Mei, and X. Chou, "Influence of Structural Parameters on Performance of SAW Resonators Based on 128° YX $LiNbO_3$ Single Crystal," *Nanomaterials*, vol. 12, no. 12, 2022, DOI: http://dx.doi.org/10.3390/nano12122109.

[23] X. Du, N. Sharma, Z. Tang, C. Leblanc, D. Jariwala, and R. H. Olsson, "Near 6-GHz Sezawa Mode Surface Acoustic Wave Resonators Using AlScN on SiC," *Journal of Microelectromechanical Systems*, vol. 33, no. 5, pp. 577–585, 2024, DOI: http://dx.doi.org/10.1109/JMEMS.2024.3430984.

[24] R. Manenti, M. J. Peterer, A. Nersisyan, E. B. Magnusson, A. Patterson, and P. J. Leek, "Surface acoustic wave resonators in the quantum regime," *Phys. Rev. B*, vol. 93, p. 041411, 2016, DOI: http://dx.doi.org/10.1103/PhysRevB.93.041411.

[25] A. Lozzi, A. De Pastina, E. T.-T. Yen, and L. G. Villanueva, "Engineered acoustic mismatch for anchor loss control in contour mode resonators," *Applied Physics Letters*, vol. 114, no. 10, 2019, DOI: http://dx.doi.org/10.1063/1.5086156.

[26] J. Segovia-Fernandez, M. Cremonesi, C. Cassella, A. Frangi, and G. Piazza, "Experimental study on the impact of anchor losses on the quality factor of contour mode AlN resonators," in *2013 Transducers & Eurosensors XXVII: The 17th International Conference on Solid-State Sensors, Actuators and Microsystems (TRANSDUCERS & EUROSENSORS XXVII)*, 2013, pp. 2473–2476, DOI: http://dx.doi.org/10.1109/Transducers.2013.6627307.

[27] S. I. Jung, C. Ryu, G. Piazza, and H. J. Kim, "A Study on the Effects of Bottom Electrode Designs on Aluminum Nitride Contour-Mode Resonators," *Micromachines*, vol. 10, no. 11, 2019, DOI: http://dx.doi.org/10.3390/mi10110758.

[28] P. Chapellier, P. Lavenus, and B. Dulmet, "A high Q 2D-length extension mode resonator for potential time-frequency applications," in *2018 Symposium on Design, Test, Integration & Packaging of MEMS and MOEMS (DTIP)*, 2018, pp. 1–5, DOI: http://dx.doi.org/10.1109/DTIP.2018.8394209.

[29] C. Xiong, X. Sun, K. Y. Fong, and H. X. Tang, "Integrated high frequency aluminum nitride optomechanical resonators," *Applied Physics Letters*, vol. 100, no. 17, p. 171111, 2012, DOI: http://dx.doi.org/10.1063/1.4707898.

[30] T. Bagci, A. Simonsen, S. Schmid, L. G. Villanueva, E. Zeuthen, J. Appel, J. M. Taylor, A. Sørensen, K. Usami, A. Schliesser *et al.*, "Optical detection of radio waves through a nanomechanical transducer," *Nature*, vol. 507, no. 7490, pp. 81–85, 2014, DOI: http://dx.doi.org/10.1038/nature13029.

[31] X.-Q. Zheng, J. Lee, and P. X.-L. Feng, "Hexagonal boron nitride nanomechanical resonators with spatially visualized motion," *Microsystems & Nanoengineering*, vol. 3, no. 1, pp. 1–8, 2017, DOI: http://dx.doi.org/10.1038/micronano.2017.38.

[32] Z. Wang, H. Jia, X. Zheng, R. Yang, Z. Wang, G. Ye, X. Chen, J. Shan, and P. X.-L. Feng, "Black phosphorus nanoelectromechanical resonators vibrating at very high frequencies," *Nanoscale*, vol. 7, no. 3, pp. 877–884, 2015, DOI: http://dx.doi.org/10.1039/C4NR04829F.

[33] J. Lee, Z. Wang, K. He, J. Shan, and P. X.-L. Feng, "High frequency $MoS_2$ nanomechanical resonators," *ACS nano*, vol. 7, no. 7, pp. 6086–6091, 2013, DOI: http://dx.doi.org/10.1021/nn4018872.

[34] A. Castellanos-Gomez, R. van Leeuwen, M. Buscema, H. S. van der Zant, G. A. Steele, and W. J. Venstra, "Single-layer $MoS_2$ mechanical resonators," *Advanced Materials*, vol. 25, no. 46, pp. 6719–6723, 2013, DOI: http://dx.doi.org/10.1002/adma.201303569.

[35] J. S. Bunch, A. M. Van Der Zande, S. S. Verbridge, I. W. Frank, D. M. Tanenbaum, J. M. Parpia, H. G. Craighead, and P. L. McEuen, "Electromechanical resonators from graphene sheets," *Science*, vol. 315, no. 5811, pp. 490–493, 2007, DOI: http://dx.doi.org/10.1126/science.1136836.

[36] M. J. Weaver, B. Pepper, F. Luna, F. M. Buters, H. J. Eerkens, G. Welker, B. Perock, K. Heeck, S. de Man, and D. Bouwmeester, "Nested trampoline resonators for optomechanics," *Applied Physics Letters*, vol. 108, no. 3, 2016, DOI: http://dx.doi.org/10.1063/1.4939828.

[37] D. Kleckner, B. Pepper, E. Jeffrey, P. Sonin, S. M. Thon, and D. Bouwmeester, "Optomechanical trampoline resonators," *Opt. Express*, vol. 19, no. 20, pp. 19 708–19 716, 2011, DOI: http://dx.doi.org/10.1364/OE.19.019708.

[38] C. Reinhardt, T. Müller, A. Bourassa, and J. C. Sankey, "Ultralow-noise SiN trampoline resonators for sensing and optomechanics," *Physical Review X*, vol. 6, no. 2, p. 021001, 2016, DOI: http://dx.doi.org/10.1103/PhysRevX.6.021001.

[39] M. J. Bereyhi, A. Beccari, R. Groth, S. A. Fedorov, A. Arabmoheghi, T. J. Kippenberg, and N. J. Engelsen, "Hierarchical tensile structures with ultralow mechanical dissipation," *Nature Communications*, vol. 13, no. 1, p. 3097, 2022, DOI: http://dx.doi.org/10.1038/s41467-022-30586-z.

[40] P. Zhang, Y. Jia, S. Yuan, M. Xie, Z. Liu, H. Jia, and R. Yang, "Unveiling the tradeoff between device scale and surface nonidealities for an optimized quality factor at room temperature in 2D $MoS_2$ nanomechanical resonators," *Microsystems & Nanoengineering*, vol. 10, no. 1, p. 140, 2024, DOI: http://dx.doi.org/10.1038/s41378-024-00763-9.

[41] R. He, X. Feng, M. Roukes, and P. Yang, "Self-transducing silicon nanowire electromechanical systems at room temperature," *Nano letters*, vol. 8, no. 6, pp. 1756–1761, 2008, DOI: http://dx.doi.org/10.1021/nl801071w.

[42] A. N. Cleland and M. L. Roukes, "Fabrication of high frequency nanometer scale mechanical resonators from bulk Si crystals," *Applied Physics Letters*, vol. 69, no. 18, pp. 2653–2655, 1996, DOI: http://dx.doi.org/10.1063/1.117548.

[43] P. Sahafi, W. Rose, A. Jordan, B. Yager, M. Piscitelli, and R. Budakian, "Ultralow dissipation patterned silicon nanowire arrays for scanning probe microscopy," *Nano letters*, vol. 20, no. 1, pp. 218–223, 2019, DOI: http://dx.doi.org/10.1021/acs.nanolett.9b03668.

[44] X. Ji, S. Roberts, M. Corato-Zanarella, and M. Lipson, "Methods to achieve ultra-high quality factor silicon nitride resonators," *APL Photonics*, vol. 6, no. 7, p. 071101, 2021, DOI: http://dx.doi.org/10.1063/5.0057881.

[45] S. Perisanu, P. Vincent, A. Ayari, M. Choueib, S. T. Purcell, M. Bechelany, and D. Cornu, "High Q factor for mechanical resonances of batch-fabricated SiC nanowires," *Applied Physics Letters*, vol. 90, no. 4, p. 043113, 2007, DOI: http://dx.doi.org/10.1063/1.2432257.

[46] J. M. Gray, C. T. Rogers, K. A. Bertness, and N. A. Sanford, "Gallium nitride nanowire electromechanical resonators with piezoresistive readout," *Journal of Vacuum Science & Technology B*, vol. 29, no. 5, p. 052001, 2011, DOI: http://dx.doi.org/10.1116/1.3622326.

[47] A. Husain, J. Hone, H. W. C. Postma, X. Huang, T. Drake, M. Barbic, A. Scherer, and M. Roukes, "Nanowire-based very-high-frequency electromechanical resonator," *Applied Physics Letters*, vol. 83, no. 6, pp. 1240–1242, 2003, DOI: http://dx.doi.org/10.1063/1.1601311.

[48] R. Gao, Z. L. Wang, Z. Bai, W. A. de Heer, L. Dai, and M. Gao, "Nanomechanics of individual carbon nanotubes from pyrolytically grown arrays," *Phys. Rev. Lett.*, vol. 85, pp. 622–625, 2000, DOI: http://dx.doi.org/10.1103/PhysRevLett.85.622.

[49] V. A. Sazonova, "A tunable carbon nanotube resonator," Ph.D. dissertation, Cornell University, 2006, DOI: http://dx.doi.org/https://hdl.handle.net/1813/3205.

[50] D. A. Walters, L. M. Ericson, M. J. Casavant, J. Liu, D. T. Colbert, K. A. Smith, and R. E. Smalley, "Elastic strain of freely suspended single-wall carbon nanotube ropes," *Applied Physics Letters*, vol. 74, no. 25, pp. 3803–3805, 1999, DOI: http://dx.doi.org/10.1063/1.124185.

[51] Y. Wang and G. J. Weng, *Electrical Conductivity of Carbon Nanotube- and Graphene-Based Nanocomposites.* Cham: Springer International Publishing, 2018, pp. 123–156, DOI: http://dx.doi.org/10.1007/978-3-319-52794-9_4.

[52] C. Yu, L. Shi, Z. Yao, D. Li, and A. Majumdar, "Thermal conductance and thermopower of an individual single-wall carbon nanotube," *Nano letters*, vol. 5, no. 9, pp. 1842–1846, 2005, DOI: http://dx.doi.org/10.1021/nl051044e.

[53] S. Schmid, L. G. Villanueva, and M. L. Roukes, *Fundamentals of Nanomechanical Resonators: Second Edition.* Springer International Publishing, 2023, DOI: http://dx.doi.org/10.1007/978-3-031-29628-4.

[54] M. Aykol, B. Hou, R. Dhall, S.-W. Chang, W. Branham, J. Qiu, and S. B. Cronin, "Clamping Instability and van der Waals Forces in Carbon Nanotube Mechanical Resonators," *Nano Lett*, vol. 14, 2014, DOI: http://dx.doi.org/10.1021/nl500096p.

[55] V. Sazonova, Y. Yaish, H. Üstünel, D. Roundy, T. A. Arias, and P. L. McEuen, "A tunable carbon nanotube electrochemical oscillator," *Nature*, vol. 431, no. 7006, pp. 284–287, 2004, DOI: http://dx.doi.org/10.1038/nature02905.

[56] B. Lassagne, D. Garcia-Sanchez, A. Aguasca, and A. Bachtold, "Ultrasensitive Mass Sensing with a Nanotube Electromechanical Resonator," *Nano Letters*, vol. 8, no. 11, pp. 3735–3738, 2008, DOI: http://dx.doi.org/10.1021/nl801982v.

[57] A. K. Hüttel, G. A. Steele, B. Witkamp, M. Poot, L. P. Kouwenhoven, and H. S. J. Van Der Zant, "Carbon nanotubes as ultrahigh quality factor mechanical resonators," *Nano Letters*, vol. 9, no. 7, pp. 2547–2552, 2009, DOI: http://dx.doi.org/10.1021/nl900612h.

[58] B. Witkamp, M. Poot, and H. S. J. Van Der Zant, "Bending-mode vibration of a suspended nanotube resonator," *Nano Letters*, vol. 6, no. 12, pp. 2904–2908, 2006, DOI: http://dx.doi.org/10.1021/nl062206p.

[59] M. Poot, B. Witkamp, M. A. Otte, and H. S. J. Van Der Zant, "Modelling suspended carbon nanotube resonators," in *Physica Status Solidi (B) Basic Research*, vol. 244, no. 11, 2007, pp. 4252–4256, DOI: http://dx.doi.org/10.1002/pssb.200776130.

[60] C. C. Wu and Z. Zhong, "Capacitive spring softening in single-walled carbon nanotube nanoelectromechanical resonators," *Nano Letters*, vol. 11, no. 4, pp. 1448–1451, 2011, DOI: http://dx.doi.org/10.1021/nl1039549.

[61] A. Eichler, J. Moser, M. Dykman, and A. Bachtold, "Symmetry breaking in a mechanical resonator made from a carbon nanotube," *Nature communications*, vol. 4, no. 1, p. 2843, 2013, DOI: http://dx.doi.org/10.1038/ncomms3843.

[62] S.-W. Lee, S. Truax, L. Yu, C. Roman, and C. Hierold, "Carbon nanotube resonators with capacitive and piezoresistive current modulation readout," *Applied Physics Letters*, vol. 103, no. 3, p. 033117, 2013, DOI: http://dx.doi.org/10.1063/1.4815988.

[63] J. Chaste, A. Eichler, J. Moser, G. Ceballos, R. Rurali, and A. Bachtold, "A nanomechanical mass sensor with yoctogram resolution," *Nature Nanotechnology*, vol. 7, no. 5, pp. 301–304, 2012, DOI: http://dx.doi.org/10.1038/nnano.2012.42.

[64] X. Wang, D. Zhu, X. Yang, L. Yuan, H. Li, J. Wang, M. Chen, G. Deng, W. Liang, Q. Li, S. Fan, G. Guo, and K. Jiang, "Stressed carbon nanotube devices for high tunability, high quality factor, single mode GHz resonators," *Nano Research*, vol. 11, pp. 5812–5822, 2018, DOI: http://dx.doi.org/10.1007/s12274-018-2085-x.

[65] P. Häkkinen, A. Isacsson, A. Savin, J. Sulkko, and P. Hakonen, "Charge sensitivity enhancement via mechanical oscillation in suspended carbon nanotube devices," *Nano Letters*, vol. 15, no. 3, pp. 1667–1672, 2015, DOI: http://dx.doi.org/10.1021/nl504282s.

[66] L. Kumar, "Energy Dissipation, Clamping and Motional Currents in Suspended Room Temperature Carbon Nanotube Resonators," Ph.D. dissertation, ETH Zurich, 2019, DOI: http://dx.doi.org/10.3929/ETHZ-B-000357371.

[67] C. C. Wu, C. H. Liu, and Z. Zhong, "One-step direct transfer of pristine single-walled carbon nanotubes for functional nanoelectronics," *Nano letters*, vol. 10, no. 3, pp. 1032–1036, 2010, DOI: http://dx.doi.org/10.1021/nl904260k.

[68] M. Muoth and C. Hierold, "Transfer of carbon nanotubes onto microactuators for hysteresis-free transistors at low thermal budget," in *2012 IEEE 25th International Conference on Micro Electro Mechanical Systems (MEMS)*, 2012, pp. 1352–1355, DOI: http://dx.doi.org/10.1109/MEMSYS.2012.6170417.

[69] A. Butzerin, S. Weikert, and K. Wegener, "Growth substrate geometry optimization for the productive mechanical dry transfer of carbon nanotubes," *Processes*, vol. 12, no. 5, 2024, DOI: http://dx.doi.org/10.3390/pr12050928.

[70] Z. Y. Ning, T. W. Shi, M. Q. Fu, Y. Guo, X. L. Wei, S. Gao, and Q. Chen, "Transversally and axially tunable carbon nanotube resonators in situ fabricated and studied inside a scanning electron microscope," *Nano Letters*, vol. 14, no. 3, pp. 1221–1227, 2014, DOI: http://dx.doi.org/10.1021/nl4040913.

[71] M. Goldsche, J. Sonntag, T. Khodkov, G. J. Verbiest, S. Reichardt, C. Neumann, T. Ouaj, N. von den Driesch, D. Buca, and C. Stampfer, "Tailoring mechanically tunable strain fields in graphene," *Nano Letters*, vol. 18, no. 3, pp. 1707–1713, 2018, DOI: http://dx.doi.org/10.1021/acs.nanolett.7b04774.

[72] M. Muoth, S.-W. Lee, and C. Hierold, "Platform for strainable, TEM-compatible, MEMS-embedded carbon nanotube transistors," in *2011 IEEE 24th International Conference on Micro Electro Mechanical Systems*, 2011, pp. 83–86, DOI: http://dx.doi.org/10.1109/MEMSYS.2011.5734367.

[73] S. Truax, S. W. Lee, M. Muoth, and C. Hierold, "Axially tunable carbon nanotube resonators using co-integrated microactuators," *Nano Letters*, vol. 14, no. 11, pp. 6092–6096, 2014, DOI: http://dx.doi.org/10.1021/nl501853w.

[74] L. Kumar, L. V. Jenni, C. I. Roman, and C. Hierold, "On-Chip Platform for Slack-Free Carbon Nanotube Resonators," *Proceedings*, vol. 1, no. 4, 2017, DOI: http://dx.doi.org/10.3390/proceedings1040374.

[75] H.-Y. Chiu, P. Hung, H. W. C. Postma, and M. Bockrath, "Atomic-Scale Mass Sensing Using Carbon Nanotube Resonators," *Nano Letters*, vol. 8, no. 12, pp. 4342–4346, 2008, DOI: http://dx.doi.org/10.1021/nl802181c.

[76] B. H. Schneider, V. Singh, W. J. Venstra, H. B. Meerwaldt, and G. A. Steele, "Observation of decoherence in a carbon nanotube mechanical resonator," *Nature Communications*, vol. 5, no. 1, pp. 1–5, 2014, DOI: http://dx.doi.org/10.1038/ncomms6819.

[77] K. Jensen, i. m. c. Girit, W. Mickelson, and A. Zettl, "Tunable nanoresonators constructed from telescoping nanotubes," *Phys. Rev. Lett.*, vol. 96, p. 215503, 2006, DOI: http://dx.doi.org/10.1103/PhysRevLett.96.215503.

[78] X. Wei, Q. Chen, S. Xu, L. Peng, and J. Zuo, "Beam to string transition of vibrating carbon nanotubes under axial tension," *Advanced Functional Materials*, vol. 19, no. 11, pp. 1753–1758, 2009, DOI: http://dx.doi.org/10.1002/adfm.200900105.

[79] M. Vollmann, C. Roman, M. Haluska, and C. Hierold, "Quantified stress relaxation in carbon nanotube resonators," in *2023 IEEE 36th International Conference on Micro Electro Mechanical Systems (MEMS)*, 2023, pp. 925–928, DOI: http://dx.doi.org/10.1109/MEMS49605.2023.10052187.

[80] M. Vollmann, C. Roman, and C. Hierold, "A phase-consistent model for the spectral transfer function of carbon nanotube resonators," in *2022 IEEE 35th International Conference on Micro Electro Mechanical Systems Conference (MEMS)*, 2022, pp. 822–825, DOI: http://dx.doi.org/10.1109/MEMS51670.2022.9699469.

[81] ——, "Tunable dual mode carbon nanotube strain gauge," in *2024 IEEE 37th International Conference on Micro Electro Mechanical Systems (MEMS)*, 2024, pp. 931–934, DOI: http://dx.doi.org/10.1109/MEMS58180.2024.10439427.

[82] A. Castellanos-Gomez, H. B. Meerwaldt, W. J. Venstra, H. S. J. Van Der Zant, and G. A. Steele, "Strong and tunable mode coupling in carbon nanotube resonators," *Physical Review B - Condensed Matter and Materials Physics*, vol. 86, no. 4, 2012, DOI: http://dx.doi.org/10.1103/PhysRevB.86.041402.

[83] A. Eichler, J. Moser, J. Chaste, M. Zdrojek, I. Wilson-Rae, and A. Bachtold, "Nonlinear damping in mechanical resonators made from carbon nanotubes and graphene," *Nature nanotechnology*, vol. 6, no. 6, pp. 339–342, 2011, DOI: http://dx.doi.org/10.1038/nnano.2011.71.

[84] Z. Ning, M. Fu, G. Wu, C. Qiu, J. Shu, Y. Guo, X. Wei, S. Gao, and Q. Chen, "Remarkable influence of slack on the vibration of a single-walled carbon nanotube resonator," *Nanoscale*, vol. 8, no. 16, pp. 8658–8665, 2016, DOI: http://dx.doi.org/10.1039/c6nr00713a.

[85] J. Moser, A. Eichler, J. Güttinger, M. I. Dykman, and A. Bachtold, "Nanotube mechanical resonators with quality factors of up to 5 million," *Nature nanotechnology*, vol. 9, no. 12, pp. 1007–1011, 2014, DOI: http://dx.doi.org/10.1038/nnano.2014.234.

[86] H. Ouakad and M. Younis, "Natural frequencies and mode shapes of initially curved carbon nanotube resonators under electric excitation," *Journal of sound and vibration*, vol. 330, pp. 3182 – 3195, 2011, DOI: http://dx.doi.org/10.1016/j.jsv.2010.12.029.

[87] T. Natsuki, N. Matsuyama, and Q.-Q. Ni, "Vibration analysis of carbon nanotube-based resonator using nonlocal elasticity theory," *Applied Physics A*, vol. 120, pp. 1309–1313, 2015, DOI: http://dx.doi.org/10.1007/s00339-015-9398-3.

[88] H. M. Ouakad, F. Najar, and O. Hattab, "Nonlinear analysis of electrically actuated carbon nanotube resonator using a novel discretization technique," *Mathematical Problems in Engineering*, vol. 2013, no. 1, p. 517695, 2013, DOI: http://dx.doi.org/10.1155/2013/517695.

[89] W. Ren, F. Li, J. Chen, S. Bai, and H. M. Cheng, "Morphology, diameter distribution and raman scattering measurements of double-walled carbon nanotubes synthesized by catalytic decomposition of methane," *Chemical Physics Letters*, vol. 359, pp. 196–202, 2002, DOI: http://dx.doi.org/10.1016/S0009-2614(02)00686-3.

[90] C. Lee, X. Wei, J. W. Kysar, and J. Hone, "Measurement of the elastic properties and intrinsic strength of monolayer graphene," *Science*, vol. 321, no. 5887, pp. 385–388, 2008, DOI: http://dx.doi.org/10.1126/science.1157996.

[91] A. Joshi, S. Harsha, and S. Sharma, "Vibration signature analysis of single walled carbon nanotube based nanomechanical sensors," *Physica E: Low-dimensional Systems*, vol. 42, pp. 2115 – 2123, 2010, DOI: http://dx.doi.org/10.1016/j.physe.2010.03.033.

[92] S. Jung, R. Hauert, M. Haluska, C. Roman, and C. Hierold, "Understanding and improving carbon nanotube-electrode contact in bottom-contacted nanotube gas sensors," *Sensors and Actuators B: Chemical*, vol. 331, p. 129406, 2021, DOI: http://dx.doi.org/10.1016/j.snb.2020.129406.

[93] A. Peigney, C. Laurent, E. Flahaut, R. Bacsa, and A. Rousset, "Specific surface area of carbon nanotubes and bundles of carbon nanotubes," *Carbon*, vol. 39, no. 4, pp. 507–514, 2001, DOI: http://dx.doi.org/10.1016/S0008-6223(00)00155-X.

[94] A. Krishnan, E. Dujardin, T. Ebbesen, P. N. Yianilos, and M. M. Treacy, "Young's modulus of single-walled nanotubes," *Physical review B*, vol. 58, no. 20, p. 14013, 1998, DOI: http://dx.doi.org/10.1103/PhysRevB.58.14013.

[95] J.-P. Salvetat, G. A. D. Briggs, J.-M. Bonard, R. R. Bacsa, A. J. Kulik, T. Stöckli, N. A. Burnham, and L. Forró, "Elastic and shear moduli of single-walled carbon nanotube ropes," *Phys. Rev. Lett.*, vol. 82, pp. 944–947, 1999, DOI: http://dx.doi.org/10.1103/PhysRevLett.82.944.

[96] M.-F. Yu, B. S. Files, S. Arepalli, and R. S. Ruoff, "Tensile loading of ropes of single wall carbon nanotubes and their mechanical properties," *Physical review letters*, vol. 84, no. 24, p. 5552, 2000, DOI: http://dx.doi.org/10.1103/PhysRevLett.84.5552.

[97] Y. Wu, M. Huang, F. Wang, X. H. Huang, S. Rosenblatt, L. Huang, H. Yan, S. P. O'Brien, J. Hone, and T. F. Heinz, "Determination of the young's modulus of structurally defined carbon nanotubes," *Nano letters*, vol. 8, no. 12, pp. 4158–4161, 2008, DOI: http://dx.doi.org/10.1021/nl801563q.

[98] A. Zandiatashbar, G.-H. Lee, S. J. An, S. Lee, N. Mathew, M. Terrones, T. Hayashi, C. R. Picu, J. Hone, and N. Koratkar, "Effect of defects on the intrinsic strength and stiffness of graphene," *Nature communications*, vol. 5, no. 1, p. 3186, 2014, DOI: http://dx.doi.org/10.1038/ncomms4186.

[99] R. Chowdhury, S. Adhikari, and J. Mitchell, "Vibrating carbon nanotube based bio-sensors," *Physica E: Low-dimensional Systems and Nanostructures*, vol. 42, no. 2, pp. 104–109, 2009, DOI: http://dx.doi.org/10.1016/j.physe.2009.09.007.

[100] H. M. Ouakad and M. I. Younis, "Nonlinear dynamics of electrically actuated carbon nanotube resonators," *Journal of Computational and Nonlinear Dynamics*, vol. 5, no. 1, p. 011009, 2009, DOI: http://dx.doi.org/10.1115/1.4000319.

[101] L. Kumar, L. V. Jenni, M. Haluska, C. Roman, and C. Hierold, "Mechanical stress relaxation in adhesively clamped carbon nanotube resonators," *AIP Advances*, vol. 8, no. 2, p. 025118, 2018, DOI: http://dx.doi.org/10.1063/1.5020704.

[102] T. Vodenitcharova and L. C. Zhang, "Effective wall thickness of a single-walled carbon nanotube," *Phys. Rev. B*, vol. 68, p. 165401, 2003, DOI: http://dx.doi.org/10.1103/PhysRevB.68.165401.

[103] B. I. Yakobson, C. J. Brabec, and J. Bernholc, "Nanomechanics of carbon tubes: Instabilities beyond linear response," *Phys. Rev. Lett.*, vol. 76, pp. 2511–2514, 1996, DOI: http://dx.doi.org/10.1103/PhysRevLett.76.2511.

[104] Y.-G. Sun, X.-H. Yao, Y.-J. Liang, and Q. Han, "Nonlocal beam model for axial buckling of carbon nanotubes with surface effect," *Europhysics Letters*, vol. 99, no. 5, p. 56007, 2012, DOI: http://dx.doi.org/10.1209/0295-5075/99/56007.

[105] Z. Xin, Z. Jianjun, and O.-Y. Zhong-can, "Strain energy and Young's modulus of single-wall carbon nanotubes calculated from electronic energy-band theory," *Phys. Rev. B*, vol. 62, pp. 13 692–13 696, 2000, DOI: http://dx.doi.org/10.1103/PhysRevB.62.13692.

[106] Z.-c. Tu and Z.-c. Ou-Yang, "Single-walled and multiwalled carbon nanotubes viewed as elastic tubes with the effective Young's moduli dependent on layer number," *Phys. Rev. B*, vol: 65, p. 233407, 2002, DOI: http://dx.doi.org/10.1103/PhysRevB.65.233407.

[107] G. Cao, X. Chen, and J. W. Kysar, "Strain sensing of carbon nanotubes: Numerical analysis of the vibrational frequency of deformed single-wall carbon nanotubes," *Physical Review B-Condensed Matter and Materials Physics*, vol. 72, no. 19, p. 195412, 2005, DOI: http://dx.doi.org/10.1103/PhysRevB.72.195412.

[108] K. N. Kudin, G. E. Scuseria, and B. I. Yakobson, "$C_2F$, BN, and C nanoshell elasticity from ab initio computations," *Phys. Rev. B*, vol. 64, p. 235406, 2001, DOI: http://dx.doi.org/10.1103/PhysRevB.64.235406.

[109] S. P. Timoshenko and J. M. Gere, *Theory of Elastic Stability*, 2nd ed. Dover Publications, 2012.

[110] J. Cai, C. Wang, T. Yu, and S. Yu, "Wall thickness of single-walled carbon nanotubes and its young's modulus," *Physica Scripta*, vol. 79, no. 2, p. 025702, 2009, DOI: http://dx.doi.org/10.1088/0031-8949/79/02/025702.

[111] A. K. Vallabhaneni, J. F. Rhoads, J. Y. Murthy, and X. Ruan, "Observation of nonclassical scaling laws in the quality factors of cantilevered carbon nanotube resonators," *Journal of Applied Physics*, vol. 110, no. 3, 2011, DOI: http://dx.doi.org/10.1063/1.3611396.

[112] H. Jiang, M.-F. Yu, B. Liu, and Y. Huang, "Intrinsic energy loss mechanisms in a cantilevered carbon nanotube beam oscillator," *Physical review letters*, vol. 93, no. 18, p. 185501, 2004, DOI: http://dx.doi.org/10.1103/PhysRevLett.93.185501.

[113] C. Urgell, W. Yang, S. L. De Bonis, C. Samanta, M. J. Esplandiu, Q. Dong, Y. Jin, and A. Bachtold, "Cooling and self-oscillation in a nanotube electromechanical resonator," *Nature Physics*, vol. 16, no. 1, pp. 32–37, 2020, DOI: http://dx.doi.org/10.1038/s41567-019-0682-6.

[114] L. G. Villanueva and S. Schmid, "Evidence of surface loss as ubiquitous limiting damping mechanism in sin micro- and nanomechanical resonators," *Phys. Rev. Lett.*, vol. 113, p. 227201, 2014, DOI: http://dx.doi.org/10.1103/PhysRevLett.113.227201.

[115] H. Jiang, M.-F. Yu, B. Liu, and Y. Huang, "Intrinsic energy loss mechanisms in a cantilevered carbon nanotube beam oscillator," *Phys. Rev. Lett.*, vol. 93, p. 185501, 2004, DOI: http://dx.doi.org/10.1103/PhysRevLett.93.185501.

[116] R. Lifshitz and M. L. Roukes, "Thermoelastic damping in micro- and nanomechanical systems," *Phys. Rev. B*, vol. 61, pp. 5600–5609, 2000, DOI: http://dx.doi.org/10.1103/PhysRevB.61.5600.

[117] M. J. Martin and B. H. Houston, "Gas damping of carbon nanotube oscillators," *Applied Physics Letters*, vol. 91, no. 10, p. 103116, 2007, DOI: http://dx.doi.org/10.1063/1.2779973.

[118] F.-H. Bao, J.-F. Bao, J. E.-Y. Lee, L.-L. Bao, M. A. Khan, X. Zhou, Q.-D. Wu, T. Zhang, and X.-S. Zhang, "Quality factor improvement of piezoelectric MEMS resonator by the conjunction of frame structure and phononic crystals," *Sensors and Actuators A: Physical*, vol. 297, p. 111541, 2019, DOI: http://dx.doi.org/10.1016/j.sna.2019.111541.

[119] L. Kumar, L. V. Jenni, M. Haluska, C. I. Roman, and C. Hierold, "Clamping and Q-Factor Improvement in a Carbon Nanotube Resonator," *Proceedings*, vol. 2, no. 13, p. 848, 2018, DOI: http://dx.doi.org/10.3390/proceedings2130848.

[120] N. J. Engelsen, A. Beccari, and T. J. Kippenberg, "Ultrahigh-quality-factor micro-and nanomechanical resonators using dissipation dilution," *Nature Nanotechnology*, vol. 19, no. 6, pp. 1–13, 2024, DOI: http://dx.doi.org/10.1038/s41565-023-01597-8.

[121] P. Avouris, Z. Chen, and V. Perebeinos, "Carbon-based electronics," *Nature nanotechnology*, vol. 2, no. 10, pp. 605–615, 2007, DOI: http://dx.doi.org/10.1038/nnano.2007.300.

[122] S. Rosenblatt, Y. Yaish, J. Park, J. Gore, V. Sazonova, and P. L. McEuen, "High performance electrolyte gated carbon nanotube transistors," *Nano letters*, vol. 2, no. 8, pp. 869–872, 2002, DOI: http://dx.doi.org/10.1021/nl025639a.

[123] S. Jung, C. Roman, and C. Hierold, "Fast nitrogen dioxide sensing with ultralow-power nanotube gas sensors," *Advanced Sensor Research*, vol. 3, no. 1, p. 2300081, 2024, DOI: http://dx.doi.org/10.1002/adsr.202300081.

[124] M. Muoth, "Clean integration of single-walled carbon nanotubes for electromechanical systems," Ph.D. dissertation, ETH Zurich, 2013, DOI: http://dx.doi.org/10.3929/ethz-a-009773215.

[125] C. Chen, V. V. Deshpande, M. Koshino, S. Lee, A. Gondarenko, A. H. MacDonald, P. Kim, and J. Hone, "Modulation of mechanical resonance by chemical potential oscillation in graphene," *Nature Physics*, vol. 12, no. 3, pp. 240–244, 2016, DOI: http://dx.doi.org/10.1038/nphys3576.

[126] L.-C. Qin, "Determination of the chiral indices (n, m) of carbon nanotubes by electron diffraction," *Physical Chemistry Chemical Physics*, vol. 9, no. 1, pp. 31–48, 2007, DOI: http://dx.doi.org/10.1039/B614121H.

[127] S. Manzeli, D. Dumcenco, G. Migliato Marega, and A. Kis, "Self-sensing, tunable monolayer mos2 nanoelectromechanical resonators," *Nature communications*, vol. 10, no. 1, p. 4831, 2019, DOI: http://dx.doi.org/10.1038/s41467-019-12795-1.

[128] C. Chen, S. Rosenblatt, K. I. Bolotin, W. Kalb, P. Kim, I. Kymissis, H. L. Stormer, T. F. Heinz, and J. Hone, "Performance of monolayer graphene nanomechanical resonators with electrical readout," *Nature nanotechnology*, vol. 4, no. 12, pp. 861–867, 2009, DOI: http://dx.doi.org/10.1038/nnano.2009.267.

[129] I. Tsioutsios, A. Tavernarakis, J. Osmond, P. Verlot, and A. Bachtold, "Real-time measurement of nanotube resonator fluctuations in an electron microscope," *Nano Letters*, vol. 17, no. 3, pp. 1748–1755, 2017, pMID: 28186773. DOI: http://dx.doi.org/10.1021/acs.nanolett.6b05065.

[130] C. T. Avedisian, R. E. Cavicchi, P. M. McEuen, X. Zhou, W. S. Hurst, and J. T. Hodges, "High temperature electrical resistance of substrate-supported single walled carbon nanotubes," *Applied Physics Letters*, vol. 93, no. 25, p. 252108, 2008, DOI: http://dx.doi.org/10.1063/1.3052867.

[131] C.-L. Zhang and H.-S. Shen, "Temperature-dependent elastic properties of single-walled carbon nanotubes: Prediction from molecular dynamics simulation," *Applied Physics Letters*, vol. 89, no. 8, p. 081904, 2006, DOI: http://dx.doi.org/10.1063/1.2336622.

[132] L. V. Jenni, "Optimization of CNT Contacts in Suspended CNTFETs and Post Dry-Transfer Processing," Ph.D. dissertation, ETH Zurich, 2020, DOI: http://dx.doi.org/10.3929/ethz-b-000394631.

[133] L. V. Jenni, L. Kumar, and C. Hierold, "Hybrid lithography based fabrication of 3D patterns by deep reactive ion etching," *Microelectronic Engineering*, vol. 209, pp. 10–15, 2019, DOI: http://dx.doi.org/10.1016/j.mee.2019.02.009.

[134] W. Kern, "The Evolution of Silicon Wafer Cleaning Technology," *Journal of The Electrochemical Society*, vol. 137, no. 6, p. 1887, 1990, DOI: http://dx.doi.org/10.1149/1.2086825.

[135] W. Zhu and E. Kaxiras, "The nature of contact between Pd leads and semiconducting carbon nanotubes," *Nano Letters*, vol. 6, no. 7, pp. 1415–1419, 2006, DOI: http://dx.doi.org/10.1021/nl0604311.

[136] L. Durrer, J. Greenwald, T. Helbling, M. Muoth, R. Riek, and C. Hierold, "Narrowing SWNT diameter distribution using size-separated ferritin-based Fe catalysts," *Nanotechnology*, vol. 20, no. 35, p. 355601, 2009, DOI: http://dx.doi.org/10.1088/0957-4484/20/35/355601.

[137] S. Jung, "Fabrication and operation of fast, ultra-low-power gas sensors with carbon nanotubes," Ph.D. dissertation, ETH Zurich, 2024, DOI: http://dx.doi.org/10.3929/ethz-b-000702474.

[138] C. C. Hu, *Modern Semiconductor Devices for Integrated Circuits - Chenming C. Hu - 9780136085256 - Electrical Engineering - Devices and Materials (134)*. Prentice Hall, 2010.

[139] S. Sze and K. N. Kwok, *Physics and Properties of Semiconductors*. John Wiley & Sons, Ltd, 2006, DOI: http://dx.doi.org/10.1002/0470068329.

[140] K. Misiakos and D. Tsamakis, "Accurate measurements of the silicon intrinsic carrier density from 78 to 340 K," *Journal of Applied Physics*, vol. 74, no. 5, pp. 3293–3297, 1993, DOI: http://dx.doi.org/10.1063/1.354551.

[141] A. Stranz, J. Kähler, A. Waag, and E. Peiner, "Thermoelectric Properties of High-Doped Silicon from Room Temperature to 900 K," *Journal of Electronic Materials*, vol. 42, no. 7, pp. 2381–2387, 2013, DOI: http://dx.doi.org/10.1007/s11664-013-2508-0.

[142] W. Fulkerson, J. P. Moore, R. K. Williams, R. S. Graves, and D. L. McElroy, "Thermal Conductivity, Electrical Resistivity, and Seebeck Coefficient of Silicon from 100 to 1300 K," *Phys. Rev.*, vol. 167, pp. 765–782, 1968, DOI: http://dx.doi.org/10.1103/PhysRev.167.765.

[143] V. M. Glazov and A. S. Pashinkin, "The thermophysical properties (heat capacity and thermal expansion) of single-crystal silicon," *High Temperature*, vol. 39, pp. 413–419, 2001, DOI: http://dx.doi.org/10.1023/A:1017562709942.

[144] M. A. Cullinan and M. L. Culpepper, "Carbon nanotubes as piezoresistive microelectromechanical sensors: Theory and experiment," *Phys. Rev. B*, vol. 82, p. 115428, 2010, DOI: http://dx.doi.org/10.1103/PhysRevB.82.115428.

[145] R. G. Knobel and A. N. Cleland, "Nanometre-scale displacement sensing using a single electron transistor," *Nature*, vol. 424, no. 6946, pp. 291–293, 2003, DOI: http://dx.doi.org/10.1038/nature01773.

[146] H. Peng, C. Chang, S. Aloni, T. Yuzvinsky, and A. Zettl, "Microwave electromechanical resonator consisting of clamped carbon nanotubes in an abacus arrangement," *Physical Review B-Condensed Matter and Materials Physics*, vol. 76, no. 3, p. 035405, 2007, DOI: http://dx.doi.org/10.1103/PhysRevB.76.035405.

[147] T. Helbling, C. Roman, L. Durrer, C. Stampfer, and C. Hierold, "Gauge Factor Tuning, Long-Term Stability, and Miniaturization of Nanoelectromechanical Carbon-Nanotube Sensors," *IEEE Transactions on Electron Devices*, vol. 58, no. 11, pp. 4053–4060, 2011, DOI: http://dx.doi.org/10.1109/TED.2011.2165544.

[148] J. Cao, Q. Wang, and H. Dai, "Electromechanical properties of metallic, quasimetallic, and semiconducting carbon nanotubes under stretching," *Phys. Rev. Lett.*, vol. 90, p. 157601, 2003, DOI: http://dx.doi.org/10.1103/PhysRevLett.90.157601.

[149] J. Wang and R. Wu, "The extended galerkin method for approximate solutions of nonlinear vibration equations," *Applied Sciences*, vol. 12, no. 6, 2022, DOI: http://dx.doi.org/10.3390/app12062979.

# Publications

## Refereed Conference Proceedings

1. M. Vollmann, "Carbon Nanotube Resonator Research", IEEE MEMS, January 2025, Kaohsiung, Taiwan
2. M. Vollmann, C. Roman, and C. Hierold, "Tunable Dual Mode Carbon Nanotube Strain Gauge", IEEE MEMS, January 2024, Austin, TX, USA
3. M. Vollmann, C. Roman, M. Haluska, and C. Hierold, "Quantified Stress Relaxation in Carbon Nanotube Resonators", IEEE MEMS, January 2023, Munich, Germany
4. M. Vollmann, C. Roman, and C. Hierold, "A Phase-Consistent Model for the Spectral Transfer Function of Carbon Nanotube Resonators", IEEE MEMS, January 2022, Tokyo, Japan

## Conference Talks

1. M. Vollmann, "Carbon Nanotube Resonator Research", IEEE MEMS, January 2025, Kaohsiung, Taiwan

## Posters

1. M. Vollmann, C. Roman, and C. Hierold, "CNT Resonators on Axial MEMS Straining Platform", 19th International Workshop on Nanomechanical Sensing, June 24-27 2024, Vienna, Austria
2. M. Vollmann, M. Basu, C. Roman, and C. Hierold, "Towards strain sensing with carbon nanotubes for monitoring respiratory activity and digital twins", Digital Twins 4 Healthcare, May 16-17 2024, Cyprus
3. M. Vollmann, C. Roman, and C. Hierold, "Tunable Dual Mode Carbon Nanotube Strain Gauge", IEEE MEMS, January 2024, Austin, TX, USA
4. M. Vollmann, C. Roman, M. Haluska, and C. Hierold, "Quantified Stress Relaxation in Carbon Nanotube Resonators", IEEE MEMS, January 2023, Munich, Germany

5. M. Vollmann, C. Roman, and C. Hierold, "A Phase-Consistent Model for the Spectral Transfer Function of Carbon Nanotube Resonators", IEEE MEMS, January 2022, Tokyo, Japan

# Student Project Supervised

1. **Lorenzo Fulciniti** *Semester project:* "Mechanical characterization of micro-sized thermal actuators "
2. **Eric Meier** *Bachelor Thesis:* "Characterization of Thermal MEMS Actuators for Straining Carbon Nanotubes"
3. **Federico De Giorgi** *Semester Project:* "Finite element modeling of distributed electrostatic forces in a CNT FET"
4. **Etienne Sciaroni** *Bachelor Thesis:* "Mechanical Cross-sensitivity Optimization of Strainable CNT Devices"
5. **Manuel Greuter** *Semester Project:* "A finite element beam model of a non-linear tunable CNT resonator"
6. **Stephan Wendelspiess** *Semester Project:* "Integration and characterization of transfer compatible chips"
7. **Marc Milewski** *Bachelor Thesis:* "Gas sensing platform"
8. **Florin Püntener** *Bachelor Studies:* "Tunable nanomechanical resonators"

# Curriculum Vitae

## Personal Details

| | |
|---|---|
| Name | Morten Vollmann |
| Birth | 16. July 1993, Berlin, Germany |
| Citizenship | Germany |

## Education

| | |
|---|---|
| 06/2019 – 03/2025 | **Micro and Nanosystems, ETH Zurich, Switzerland**<br>PhD Dissertation "Combination of Axial Strain Tuning and Gate Bias Tuning for Carbon Nano Tube Resonators" |
| 09/2016 – 10/2018 | **Department of Mechanical Engineering, EPF Lausanne, Switzerland**<br>MSc in Mechanical Engineering |
| 10/2012 – 09/2016 | **TU Berlin, Germany**<br>BSc in Engineering Science |

## Work Experience

| | |
|---|---|
| 06/2019 – 03/2025 | **Micro and Nanosystems, ETH Zurich, Switzerland**<br>Research and teaching assistant |
| 03/2018 – 10/2018 | **IBM Research Rueschlikon, Switzerland**<br>Research intern |
| 10/2014 – 09/2016 | **TU Berlin, Berlin, Germany**<br>Teaching assistant |
| 03/2015 – 10/2015 | **The College of New Jersey, Ewing, NJ, USA**<br>Research scientist, optical laboratory responsible |
| 08/2012 – 09/2012 | **Melag Medizintechnik, Berlin, Germany**<br>Intern |

## Languages

| | |
|---|---|
| **German** | Native |
| **English** | Advanced |
| **French** | Intermediate |

**Scientific Reports on Micro and Nanosystems**

**edited by Prof. Dr. Christofer Hierold**
**ETH Zürich Micro and Nanosystems**

Vol. 28: Verena Maiwald, **A Microelectromechanical Switch for Bandpass Vibration Detection.**
1st Edition 2018. XVI, 164 pages. € 64,00. ISBN 978-3-86628-614-6

Vol. 29: Michelle Müller, **Micromechanical broadband vibration amplitude-amplifier for microseismic and acoustic emission detection.**
1st Edition 2019. XVIII, 168 pages. € 64,00. ISBN 978-3-86628-627-6

Vol. 30: Silvan Marc Staufert, **Conformal Parylene-C Media Separating Membranes for Pressure Sensing Applications in Ventricular Assist Devices.**
1st Edition 2019. XVIII, 150 pages. € 64,00. ISBN 978-3-86628-636-8

Vol. 31: Lalit Kumar, **Energy Dissipation, Clamping and Motional Currents in Suspended Room Temperature Carbon Nanotube Resonators.**
1st Edition 2019. XXIV, 182 pages. € 64,00. ISBN 978-3-86628-650-4

Vol. 32: Sebastian Eberle, **Ultra-clean suspended carbon nanotube gas sensors - concept for large scale fabrication and sensor characterization.**
1st Edition 2019. XX, 180 pages. € 64,00. ISBN 978-3-86628-659-7

Vol. 33: Laura Vera Jenni, **Optimization of CNT Contacts in Suspended CNTFETs and Post Dry-Transfer Processing**.
1st Edition 2019. XXII, 170 pages. € 64,00. ISBN 978-3-86628-660-3

Vol. 34: Ian Mihailovic, **Ag/BiSe memristors for sensor data storage: a novel concept for zero-power sense-log devices.**
1st Edition 2022. XXII, 146 pages. € 64,00. ISBN 978-3-86628-769-3

Vol. 35: Stefan Nedelcu, **Energy efficient analog mixed-signal front ends for CNT-FET NO2 air-quality nanosensors.**
1st Edition 2022. XVI, 232 pages. € 64,00. ISBN 978-3-86628-779-2

Vol. 36: Johannes Weichart, **Artificial Fingertip with Embedded High Resolution Tactile Sensing.**
1st Edition 2023. (10), XX, 158 pages. € 64,00. ISBN 978-3-86628-802-7

Vol. 37: Katrina Klösel, **Multifunctional materials: exploiting the versatility of $Bi_2Se_3$ for multimodal sensing and zero power sensor systems.**
1st Edition 2024. (8), XX, 188 pages. € 64,00. ISBN 978-3-86628-816-4

**Scientific Reports on Micro and Nanosystems**

**edited by Prof. Dr. Christofer Hierold**
**ETH Zürich Micro and Nanosystems**

Vol. 38: Matthias Alexander Dupuch, **Implantable Pressure Sensor Encapsulation for Ventricular Assist Device Control.**
1st Edition 2024. (8), XXII, 162 pages. € 64,00. ISBN 978-3-86628-834-8

Vol. 39: Signe Lin Kuei Vehusheia, **Exploring heat flux as a marker for the early detection of infection.**
1st Edition 2024. (8), XXXII, 214 pages. € 64,00. ISBN 978-3-86628-835-5

Vol. 40: Seoho Jung, **Fabrication and Operation of Fast, Ultra-low-power Gas Sensors with Carbon Nanotubes.**
1st Edition 2024. (8), XXVIII, 178 pages. € 64,00. ISBN 978-3-86628-836-2

Vol. 41: Lucrezia Maini, **Exploratory Investigation of Ultrasound Interrogated Passive Sensors based on an Acoustic Meta-material.**
1st Edition 2025. (8), XXXIV, 270 pages. € 98,00. ISBN 978-3-86628-845-4

**Hartung-Gorre Verlag, Konstanz http://www.hartung-gorre.de**